DISPOSITIONS

Dispositions

STEPHEN MUMFORD

OXFORD UNIVERSITY PRESS
1998

Oxford University Press, Great Clarendon Street, Oxford OX2 6DP
Oxford New York
Athens Auckland Bangkok Bogota Bombay Buenos Aires
Calcutta Cape Town Dar es Salaam Delhi Florence Hong Kong Istanbul
Karachi Kuala Lumpur Madras Madrid Melbourne Mexico City
Nairobi Paris Singapore Taipei Tokyo Toronto Warsaw
and associated companies in
Berlin Ibadan

Published in the United States
by Oxford University Press Inc., New York

First published 1998

British Library Cataloguing in Publication Data
Data available

Library of Congress Cataloging in Publication Data
Mumford, Stephen.
Dispositions/Stephen Mumford.
Includes bibliographical references and index.
1. Dispositions (Philosophy) I. Title.
BD374.M85 1998 111–dc21 98–3549
ISBN 0-19-823611-5

1 3 5 7 9 10 8 6 4 2

Typeset by J&L Composition Ltd, Filey, North Yorkshire
Printed in Great Britain
on acid-free paper by
Bookcraft (Bath) Ltd,
Midsomer Norton, Somerset

PREFACE

Commonplace dispositions are the elasticity of a rubber band, the fragility of a wineglass, and the solubility of sugar and salt. Such dispositions are to be found in abundance. A class of predicates picks them out. It is not just predicates with the tell-tale suffixes that do this; arguably 'soft', 'opaque', 'sweet', and 'poisonous' are indicative of the dispositional as much as 'flexible' and 'edible'. In addition to these terms for physical dispositional properties there is an established tradition that treats certain mental terms as indicating dispositions rather than occurrent states. Thus 'trust-worthy', 'believes . . .', 'expects . . .', and 'loves . . .' are arguably to be understood as attributions of dispositions rather than descriptions of present mental occurrences.

A number of general philosophical questions are raised by our use of a dispositional vocabulary. Do these terms name real dispositional properties or is their true meaning more complex? Some have suggested that dispositions are not properties at all and that the dispositional vocabulary has value only insofar as it refers to past actual, or future possible, events. Is a disposition a cause of its manifestation? Prima facie it seems right to say that it is but it emerges that whether or not dispositions are fit to be causes of anything is dependent upon a theory of what a disposition consists in. A further question concerns why things have the dispositions they have and not others. Some strategies suggest that there are reasons for this. In contrast, some empiricist treatments render the possession of dispositions mysterious and our beliefs about their possession justified only inductively.

Some dispositions are a special cause of puzzlement. A common view is that underlying each disposition is a structure consisting of components which are responsible for that disposition. This gives us a structural analysis as part of a physical theory of dispositions. These components themselves have dispositions, however. In turn this suggests that the components have smaller components with even more dispositions. Can these substructures go on forever, with ever smaller components and ever more dispositions, or does the analysis come to an end somewhere? What could end the analysis? One suggestion is that it ends with dispositions that are ungrounded in any structural components. There is an understanding of modern physics that it makes an appeal to these dispositions as the ultimate basis of the physical realm. Such dispositions are the forces, fields, propensities, and potentials that exist at a subatomic level and arguably there are no other properties underlying these that explain their presence. This kind of disposition is puzzling because the reasons given for the presence of most dispositions cannot be given for these. Evidently there are two types of disposition. This is not the only distinction that can be drawn among dispositions, though; other differences can be found. The aim of this study is not just to note differences, however. There is also an attempt at a general account, for there ought to be some reason why all these different things are understood as dispositions. Can we find some principled basis on which evidently differing phenomena are classed under the same genus?

I am looking for a credible account of where dispositions fit into the world. If we treat them as nothing more than possibilities, then, in our world of actuality, there seems to be no place for them. Dispositions should correctly be understood as more than mere possibilities, however. To say something has a disposition is to say something about how it is actually. What is clearly important is that if we are to have a tenable theory of dispositions we had better start with a correct analysis of what is meant by a disposition ascription. My approach will thus begin with an analysis of the dispositional–categorical distinction. This is traditionally either taken for granted or collapsed altogether. It is a distinction that I feel has not yet been defensibly stated but which, ultimately, I think is sustainable. Only after this distinction is understood can we progress to the ontological question of where dispositional and categorical properties fit into the world. One theory I will be arguing against divides the world in two insofar as it understands things to be composed of two fundamentally different types of property: the dispositional and the categorical. This bifurcation of reality is costly but also unfounded and avoidable. An alternative is plausible. The dispositional and the categorical do not ascribe two different types of property that objects or kinds possess; rather, they are two modes of denoting the very same properties of these things. When we think of dispositions in the context of the surrounding issues of causation, explanation, and manifestation, this view becomes the most compelling. What we end up with is a developed account of what the dispositional idiom is and what we use it for. Disposition ascriptions are, on this view, units of explanation which do not provide a full explanation but nevertheless have a valuable explanatory role to play. This study is thus, in a tradition begun by D. H. Mellor, a defence of dispositions.

I want to thank a number of people who in various ways have helped to bring this book about. First, I want to acknowledge a debt to those who have given me philosophical input on these issues or have given me comments on preliminary drafts. It has delighted me that so many people have freely given their time and have had the interest to read my work and discuss the ideas with me. I acknowledge specific intellectual debts in this respect to Robert Black, Nancy Cartwright, John Divers, Robert Kirk, Robin Le Poidevin, David Lewis, Harry Lewis, Jonathan Lowe, Duncan MacFarland, Charlie Martin, Hugh Mellor, Alexander Miller, Ullin Place, and Andrew Wright. I have also received useful comments when presenting papers in seminars at Leeds, Nottingham, Birkbeck, Hertfordshire, Kent, and Birmingham. I offer my sincere thanks to the anonymous Oxford University Press reader who took the time to make detailed comments on the first draft of this book. These comments were extremely valuable and I have tried to do them justice. I would like to thank my Department at Nottingham, and all my colleagues there, for facilitating a healthy research environment that has been conducive to finishing a lengthy project such as this. Also I am grateful to the editors of *Philosophical Quarterly* and *Dialectica* for permission to use material that appears in Chapters 4 and 10 respectively.

Finally, I must thank all those who have given me personal support over the years in which I have been working towards completion of this book. I have had encouragement and practical support from many people including my wife Maggie, to whom this book is dedicated, my late father, my family, and John Divers, Alexander Miller, Robin Le Poidevin, and Eros Corazza. Without such help I could not have got anywhere near producing this work which is a testament to the kindness I have received.

CONTENTS

I

Threats and Promises

1.1 *A World of Threats and Promises*

If we want to entertain seriously the view that the world contains dispositional properties, we have much work to do. Dispositions are objects of suspicion. The first task is to understand why. Nelson Goodman has provided some reasons. He says:

> Besides the observable properties it exhibits and the actual processes it undergoes, a thing is full of threats and promises. The dispositions or capacities of a thing—its flexibility, its inflammability, its solubility—are no less important to us than its overt behaviour, but they strike us by comparison as rather ethereal. And so we are moved to inquire whether we can bring them down to earth; whether, that is, we can explain disposition-terms without any reference to occult powers.[1]

It is this problem that is the chief concern of the present work. It is a problem acknowledged ever since the Aristotelian world-picture of 'real potentialities' came into conflict with its mechanistic rival. More recently it is the mechanistic view that has been thrown into crisis. As a result, the nature of dispositions has become an even more urgent concern. I aim to bring dispositions down to earth in the way Goodman wants and, in so doing, make appeal to them legitimate and scientific.

An initial characterization of dispositions is needed. Goodman suffices as a starting point for he makes three remarks on what is typically meant by such 'threats and promises'.

First, Goodman notes that they are possessed by 'things'. We can say that a particular wineglass is fragile or that a particular rubber band is elastic. These are attributions of dispositions to individual objects. Such attributions are commonplace and, for the

[1] N. Goodman, *Fact, Fiction, and Forecast* (Indianapolis, 1955), 40.

most part, regarded as unproblematic by those who make them. I refer to such an attribution as a *disposition ascription*.

Objects are not the only things that we take to be possessors of dispositions. We ascribe dispositions to types of thing, as when we say that sugar and salt are soluble, or that a certain chemical is volatile, or that tin is malleable. Here dispositions are being ascribed to kinds. Sugar, the whole sugar-kind, is a soluble kind. This disposition ascription goes beyond an ascription to any particular sugarcube or sample of sugar. Indeed, we could still say that sugar was a soluble kind even if all samples of the kind were destroyed. Chemists make disposition ascriptions to theoretical elements, towards the end of the periodic table, of which no sample is known to exist.

The final distinguishable category of things to which dispositions are ascribed is persons. Persons are ascribed two types of disposition, corresponding to the distinction Strawson drew between M-predicates and P-predicates.[2] The former type of disposition ascription is one that could be ascribed also to an inanimate object; for instance, persons may be said to be insoluble, soft, and opaque. These disposition ascriptions apply in virtue of people having physical bodies. Many disposition terms are reserved, however, for ascription solely to things with minds. We say that persons are generous or moody or that they are good musicians or bad carpenters.

Some mental attribute terms are clearly dispositional but are they all of such a nature? Consider what it means to say that *x* loves *y*. Wittgenstein remarked that love was not a feeling but, rather, something that was put to the test.[3] We should not interpret an ascription of love, on this view, as an ascription of an occurrence, or current (mental) event, for we can rightly say that someone is in love even though they are asleep. To say that someone is in love is to say that they are disposed to behave in certain ways in a variety of circumstances. They are disposed to buy flowers if passing a florist or to write sentimental poetry when otherwise unoccupied. The most famous such account of mental dispositions is Ryle's.[4] Ryle suggested that *all* mental ascriptions were disposition

[2] P. F. Strawson, *Individuals* (London, 1959), 104.

[3] L. Wittgenstein, *Remarks on the Philosophy of Psychology*, i (Oxford, 1980), sects. 115, 959.

[4] G. Ryle, *The Concept of Mind* (London, 1959).

ascriptions. He had a specific purpose in mind: to destroy the 'myth of the ghost in the machine' by explaining away the need for mysterious, hidden, mental occurrences. Ryle's analysis amounts to an important challenge to the Cartesian view. It suggests that mental predications do not refer to occurrences in the head but, instead, are disposition ascriptions.

In summary, dispositions are ascribed to at least three distinguishable classes of things: objects, substances, and persons. I mention this to demonstrate the significance and pervasive nature of dispositional discourse. Talk and attribution of dispositions is a widely applicable discourse. But if dispositions are problematic or ethereal, as Goodman suggests, then any discourse that uses disposition terms is equally problematic.

So much for the subjects of disposition ascription. What, I now consider, is it that we ascribe to such particulars and kinds when we make a disposition ascription? The most natural answer is that it is properties that are ascribed. Dispositions are ordinarily understood as properties of objects and kinds. This is where Goodman's second point can be brought to bear: the dispositions of a thing are as important to us as any other of its properties. Knowledge of the dispositions of a thing is knowledge of how it can or will behave in various circumstances, hence Goodman's characterization of dispositions as 'threats and promises'. Our treatment of the things around us is shaped by our knowledge of their dispositions to behave. The disposition of wine to smell and taste a certain way, to produce a certain sensation on the tongue, and to intoxicate are its most important properties for us. The disposition of certain substances to emit radiation has an obvious overriding importance and determines crucially the way we treat those things. Further, we may take some of the essential properties of things to be dispositional, as when we characterize copper with reference to the properties ductile, strong, hard, malleable, fuses with difficulty, forms alloys, rusts less than iron, and so on.[5] It is clear, therefore, that we do take seriously the dispositional properties of a thing.

However, the view that dispositions are properties comes under a number of attacks. Besides the special question of whether mental dispositions are dispositional *properties*, which may be just a

[5] R. Harré and E. H. Madden, *Causal Powers* (Oxford, 1975), 24. See also C. McGinn, 'A Note on the Essence of Natural Kinds', *Analysis*, 35 (1975).

dispute about ordinary usage,[6] there are two other general ways in which dispositions have been denied the status of properties. The first is the Ryle–Wittgenstein view that to ascribe a disposition is not to ascribe a property but merely to say how something will behave in certain circumstances. Dispositional properties are, on this view, logical fictions; what really exist are regularities of events. The second attack takes dispositions to be primitive, unsophisticated, pre-scientific terms which, like 'phlogiston', are explained away with the advances of science. Disposition terms are, on this view, just place-holders: terms we use when we are ignorant of the supposedly real properties of a thing such as its height, weight, shape, and molecular structure. I shall be arguing that we should be sceptical about these views which challenge the status of dispositions.

This brings us to Goodman's third observation. Dispositions are considered ethereal: properties that somehow are not always manifest but which seem to lurk in a mysterious realm intermediate between potentiality and actuality. It is this feature that challenges us to bring dispositions down to earth and account for them in terms of what is actual.

Dispositional properties are typically contrasted with 'categorical' properties such as shape and structure and 'categorical' and 'non-dispositional' are usually interchangeable. This raises another problem, for any account of dispositions, concerning how we are to distinguish the dispositional from the categorical. Goodman seems to suggest that the non-dispositional properties of things are their observable properties but this will not do. Evidently some categorical properties are unobservable too. Molecular structure is traditionally conceived of as a categorical property but a molecular

[6] In the case of mental dispositions it seems odd to speak of a mental dispositional *property*, whereas we are quite happy to speak of physical dispositions as properties. What seems most natural is to speak of mental *dispositions* and non-mental *dispositional properties*; certainly philosophers have adopted such a practice in the literature. What counts, if such a distinction is to be taken seriously, is whether we can outline a significant difference in kind between mental and non-mental dispositions that warrants the one kind being classed as properties and the other not. If there is no warrant, then the refusal to speak of mental dispositional *properties* can be dismissed. Such a difference between dispositions and dispositional properties, I maintain, must be found, rather than stipulated, as Bergmann attempts (G. Bergmann, 'Dispositional Properties and Dispositions', *Philosophical Studies*, 6 (1955)), for such a stipulation would seem to have no relevance to the actual usage of such terms.

structure is too small to be observed. If we are going to make a clear distinction between the dispositional and the non-dispositional, we had better find firmer grounds. More promising is Goodman's characterization of dispositions as 'threats and promises'. To ascribe a disposition is to suggest possibilities of behaviour. It is to say that something could or would happen if the circumstances were right.

It is clear that there are a number of preliminary questions that must be answered before we can become immersed in the metaphysical issues. These questions are so often overlooked but are, I shall argue, prior to the ontological ones. The preliminary questions I will try to answer are about what dispositions are contrasted with and whether there are sustainable grounds for the contrast.

1.2 *Five Dispositions*

In this section I consider five examples of dispositions. These show the diversity and flexible application of disposition concepts. They will also serve to introduce some of the philosophical issues central to an understanding of dispositions.

(i) *Fragility*

One of the most typical of 'philosophers' dispositions' is fragility which is ascribed to objects such as vases and panes of glass.[7] Ascriptions of fragility are often warnings for they are understood to mean that breakage is easily brought about. An accidental dropping or knocking is very likely to result in smashing. It seems that although virtually all objects are breakable, it is only those which break easily, under very little force of impact and in a wide variety of circumstances, that are called fragile.

This shows a key feature of dispositions. To each disposition there corresponds a typical manifestation but a disposition ascription can be true though no manifestation occurs. In the case of

[7] The term 'philosophers' disposition' is used by E. Prior, *Dispositions* (Aberdeen, 1985), 1, to designate those paradigm cases most often discussed by the philosophical community.

fragility, the typical manifestation is breakage. In the case of solubility, the typical manifestation is dissolving. Some dispositions are 'multiply manifested', meaning that they have a variety of possible manifestations. The multiple manifestations of elasticity include stretching, contracting, bouncing, deforming, and reforming. None of these events need ever occur for a disposition ascription to be true. A vase may be fragile and may have been so for centuries though it has never manifested its fragility in an actual breakage. So where does its fragility reside? In what fact of the matter does the fragility of the vase consist? We probably don't want to say that it consists in the possession of some 'occult' power that lurks in a halfway realm between potentiality and actuality. But then, what do we want to say?

There is no commitment to actual breakage, hence some notion of conditionality is invoked. What kind of conditionality is relevant is controversial. Mellor has commented that we shouldn't assume that the conditional involved is always counterfactual; for we do not want to deny that something has a particular disposition which it is actually manifesting at present.[8] In other cases, the conditional *will* be counterfactual, however. Once a fragile vase manifests its fragility in breakage its fragility is lost; for there is no longer the possibility of that object manifesting that disposition. Similarly, when sugar is dissolved it is wrong still to call it soluble, for it makes no sense to say that what remains can be further dissolved. Making a general rule about the type of conditional involved in a disposition ascription seems unnecessarily restrictive, therefore, but something more precise should be said about the relation between disposition ascriptions and conditionals. I examine this issue in Chapters 3 and 4.

A further key notion raised by the existence of unmanifested dispositions is that of stimulus conditions. These conditions provide the content of the 'if . . .' clauses or antecedents of the conditionals connected with disposition ascriptions. A fragile object would break easily if certain events occurred, such as the vase being dropped or it being knocked. The antecedent events are the stimuli for the disposition. This raises the question of the relation between the stimulus and the disposition in producing

[8] D. H. Mellor, 'In Defense of Dispositions', *Philosophical Review*, 83 (1974), 169.

the manifestation. The correct way to relate the concepts of disposition, manifestation, conditional, and stimulus will be an essential part of the investigation.

(ii) *Belief*

Occurrent and dispositional senses of belief can be distinguished.[9] Occurrent beliefs are mental events, such as John's belief at 3 o'clock on Thursday that he is being watched. Dispositional beliefs are more enduring states that can be ascribed over longer periods of time and need not be currently entertained for their ascription to be true. There are disagreements over exactly how much of our belief is occurrent and how much is dispositional.[10] Candidate dispositional beliefs are the belief that eight eights are sixty-four, the belief that zebras have black and white stripes, and the belief that the chair on which you are sitting will support your weight.

Notwithstanding the obvious difference from the disposition of fragility—that beliefs are ascribed to persons rather than objects—there are enough similarities with paradigm cases of dispositions. We say that someone believes that eight eights are sixty-four because that would be the response they would most frequently give if questioned on the matter. These beliefs are dispositional in the sense that they have typical manifestations, in verbal behaviour, typical stimuli, in questions, and they can be truthfully ascribed even though there is no current manifestation.

The difference between these cases and dispositional beliefs such as that the chair will support one's weight is that the behaviour in which the latter beliefs are manifested is typically non-verbal. Even if they do not say so verbally, we can ascribe the belief to *x* purely on the basis that *x* sits upon the chair in question and shows no sign in behaviour that they are bracing themselves for a fall or rehearsing a stunt as part of a clown act.

[9] See I. Levi and S. Morgenbesser, 'Belief and Disposition', *American Philosophical Quarterly*, 1 (1964) and D. M. Armstrong, *Belief, Truth and Knowledge* (Cambridge, 1973), ch. 2.

[10] Instrumentalism and Interpretationism will emphasize a dispositional approach to belief more so than a Cartesian or Lockean realist position. See Dennett, 'Intentional Systems', in *Brainstorms* (Montgomery, Vt., 1978), for accounts of these positions.

(iii) *Bravery*

The case of bravery provides us with an example of a disposition which is a complex of subdispositions. One necessary condition of bravery, for instance, is a disposition to perceive certain situations as fearful. As Nicias says in the *Laches*, we do not call animals or children brave if they act out of ignorance of danger.[11] Their act could be behaviourally similar to the act of a brave person in similar circumstances but because they have no fear of those circumstances they are fearless without being brave. Bravery is a complex of (at least) two subdispositions: a particular disposition to represent certain but not all situations as fearful and a disposition to the right response towards such fear. Butler has suggested that most character trait dispositions have similar complex natures and that we do not do them full justice if we try to understand them in terms of a single conditional.[12]

It may be thought that character traits are significantly different from the paradigm physical dispositions of objects in that it makes no sense to ascribe bravery to someone who has never acted bravely although we can ascribe fragility to an object that has never smashed. This raises the difficult question of the truth-conditions of disposition ascriptions but a case could be made for saying that someone is brave even though they have never acted bravely. There may have been no appropriate circumstances in which they could have acted bravely, or if they were in such a situation they may have been drunk or affected by food additives.[13] We must admit the possibility of such cases but then the question returns of what such a person's bravery consists in. Is there a fact of the matter in this case?

(iv) *Thermostats*

The essence of many artefacts is dispositional. Thermostats, thermometers, axes, spoons, and batteries have dispositional essences. What it is that makes certain artefacts the thing that they are is

[11] Plato, *Laches*, 196c–7e.

[12] D. Butler, 'Character Traits in Explanation', *Philosophy and Phenomenological Research*, 49 (1988).

[13] A. Wright, 'Dispositions, Anti-Realism and Empiricism', *Proceedings of the Aristotelian Society*, 91 (1990/1), 48.

that they have a particular set of dispositions. The exact mechanisms that equip a thermostat with the dispositions it has are, in a significant way, irrelevant to the fact that it is a thermostat. Two thermostats may be constructed quite differently. The exact mechanism is inessential insofar as something could still be a thermostat if it had a different constitution so long as it was equipped, in some way, with the right sort of dispositions for a thermostat.

What is the right sort of disposition for a thermostat? Roughly, for something to be a thermostat it must be sensitive to changes in temperature and be able to trigger a switch if a pre-calibrated temperature threshold is crossed. Anything which has this disposition, if I am analysing the concept correctly, is a thermostat regardless of the constitution that affords such an ability.

This example has introduced an important distinction: the distinction between a disposition and a mechanism that somehow constitutes it or endows it. The exact nature of this distinction and the relation between the notions of disposition and underlying mechanism will be one of my main concerns. It will lead to a number of important questions. Need all dispositions be supported by a mechanism? Do mechanisms cause dispositions, constitute them, or is some other relation involved? Can we say that the disposition and mechanism are two distinct qualities and does this mean that the world is populated by two very different types of property?

(v) *Divisibility by 2*

There is a different kind of disposition to those discussed so far. Many of the properties of abstract objects, such as numbers and geometrical shapes, could be considered dispositional. This was a fact recognized by Aristotle[14] but ignored since. One such property is being divisible by 2. The number 86 has it but the number 53 does not. It is a property that is possessed by (approximately) half of the natural numbers.

I will make a preliminary distinction between 'concrete' and 'abstract' dispositions, the latter being those dispositions possessed by abstract entities such as numbers and geometrical shapes. The

[14] Aristotle, *Metaphysics*, 1019b33. Aristotle's example is of 'potency in geometry'.

basis for making a distinction is that the dispositions of abstract entities have no causal powers and produce no change. Aristotle made basically the same distinction when he said:

A 'potency' or 'power' in geometry is so-called by a change of meaning. These senses of 'capable' or 'possible' involve no reference to potency. But the senses which involve reference to potency all refer to the primary kind of potency; and this is a source of change in another thing or in the same thing *qua* other.[15]

To Aristotle's primary kind of potency corresponds a concrete disposition which has causal significance in a way to be explained in Chapter 6; to the powers in geometry correspond abstract dispositions which have no such causal significance.[16] I will be presenting an account of dispositions which will explain why abstract dispositions may with justification be regarded as dispositional.

It may be wondered why the examples I have presented are classed specifically as dispositions and not as members of other related classes. Related classes include tendencies, capacities and incapacities, powers and forces, potentialities and propensities, abilities and liabilities.

I take it that the dispositional is a genus that can accommodate these subclasses as species. The belief that eight eights are sixty-four may be a capacity. Fragility may be a liability. Bravery may be a tendency. We can grant these classifications. Of the set of related terms, some may be taken as equivalent to dispositions, such as powers and potentialities. Other terms such as tendency, ability, and propensity may refer to dispositions of a particular kind. Champlin explicates the meaning of 'tendency', for instance, showing how the term contains an ambiguity not present in 'disposition'.[17] Ryle makes the simple distinction that an ability is a

[15] *Metaphysics*, 1019b33–20a1.

[16] Elsewhere Aristotle develops his point: '"potency" and the word "can" have several senses. Of these we may neglect all the potencies that are so called by an equivocation. For some are called so by analogy, as in geometry we say one thing is or is not a "power" of another by virtue of the presence or absence of some relation between them. But all potencies that conform to the same type are originative sources of some kind, and are called potencies in reference to one primary kind of potency which is an originative source of change in another thing or in the thing itself *qua* other' (*Metaphysics*, 1046a4–12).

[17] T. S. Champlin, 'Tendencies', *Proceedings of the Aristotelian Society*, 91 (1990/1).

(mental) disposition that it is useful to have, a liability is a disposition that it is a hindrance to have.[18] 'Propensity' is a term which is often used in scientific literature to denote a disposition that can be probabilistic rather than 'sure-fire'.[19]

If we can show that these related notions are all either equivalent to dispositions or types of disposition, then the investigation need not be hampered by considering each subcategory. The notion of a disposition raises enough philosophical problems. If the treatment of dispositions is successful, however, the general account should be applicable to the subcategories with qualifications where necessary.

1.3 *Dispositional Explanations*

Dispositions are posited as explanations of past events and grounds for the prediction of future events. What exactly their role is in the production of such events and whether they have any role at all to play in explanation are, however, controversial points. If dispositions have no explanatory role, as some would argue, then there would seem to be no use in their ascription.

Set against the stance which is dismissive of their explanatory role are two views. First, there is the view that dispositions have an explanatory role but that it is a merely heuristic one. The making of a disposition ascription is akin to leaving a promissory note that states that a further explanation is available. The second view is that a dispositional explanation can be a final explanation that ends a chain of explanation and can be explicated no further.

How is it that the ascription of a disposition would provide an explanation of an event? Hempel says that dispositional explanation is a theoretical explanation in the deductive-nomological form.[20] An example of the form of a dispositional explanation is given as:

[18] *The Concept of Mind*, 130–1. Ryle speaks of capacities as competences and liabilities as limitations.

[19] K. R. Popper, 'The Propensity Theory of the Calculus of Probability, and the Quantum Theory', in S. Körner (ed.), *Observation and Interpretation* (London, 1957).

[20] C. G. Hempel, 'Dispositional Explanation', in R. Tuomela (ed.), *Dispositions* (Dordrecht, 1978). For the exposition of deductive-nomological explanation see C. G. Hempel, 'Explanation in Science and in History', in R. G. Colodny (ed.), *Frontiers of Science and Philosophy* (London, 1962). See also H. B. Dalrymple, 'Dispositional and Causal Explanation', *South-West Journal of Philosophy*, 6 (1975).

C_1: *i* was in a situation of kind *S*
C_2: *i* has the property M
L: Any *x* with property M will, in a situation of kind *S*, behave in manner *R*
Therefore,
E: i behaved in manner *R*,

where C_1 and C_2 are initial conditions, *L* is a general law, and together C_1, C_2, and *L* entail *E*. Deductive-nomological explanation in general has come in for criticism, however. That it is explanatory at all is attacked by McMullin who would put in its place a 'hypothetico-structural' form of explanation.[21] This explains why *x* φ-ed at time *t* by ascribing a structure to *x* which is responsible for its φ-ing. Such explanation is preferable, McMullin tells us, for two reasons: first that it offers an explanation of why something exhibits regularity of behaviour, rather than merely stating that there is such a regularity and, second, it introduces conceptual novelty, making progress by opening up a hitherto hidden world of processes and structures.

The explanation of dispositions by structures is, of course, no new development and exemplifies a well-entrenched desire among philosophers to replace appeal to dispositions in explanation. Boyle, who was one of the first to give voice to this desire, tells us that 'the *Corpuscularians* will show that the very qualities of this or that ingredient flow from its particular texture and the mechanical affections of the corpuscles it is made up of'.[22] All the mysterious powers in philosophy, passed down from Aristotle, can be explained in terms of shapes and structures; that is, the primary qualities of things. For example: 'the solidity, taste, &c., of salt may be fairly accounted for by the stiffness, sharpness, and other mechanical affections, of the minute particles whereof salt consists',[23] or the power of a lock to be opened by a key can be explained in terms of the shape of the lock and the shape of the key.[24] Similarly, 'gunpowder itself owes its aptness to be fired and

[21] E. McMullin, 'Structural Explanation', *American Philosophical Quarterly*, 15 (1978).

[22] R. Boyle, 'About the Excellency and Grounds of the Mechanical Hypothesis' (1647), in M. A. Stewart (ed.), *Selected Philosophical Papers of Robert Boyle* (Manchester, 1979), 151.

[23] Ibid. 149.

[24] R. Boyle, 'The Origin and Forms of Qualities' (1666), in M. A. Stewart (ed.), *Selected Philosophical Papers of Robert Boyle*, 23.

exploded to the mechanical contexture of more simple portions of matter—*nitre*, *charcoal*, and *sulphur*.[25] Boyle goes into detail on the power of beaten glass to poison. Though the power to poison is regarded as a real quality of the glass it 'is really nothing distinct from [the primary qualities which compose] the glass itself'.[26] We merely state the mechanism by which the substance produces its effect: small, glassy fragments, being 'many [and] rigid . . . and endowed with sharp points and cutting edges . . . pierce or wound the tender membranes of the stomach and guts, . . . whereby naturally ensue great gripings and contortions of the injured parts'.[27] Further, ground glass is no poison to some animals because their guts 'are usually lined with a slimy substance' which sheaths the minute powders.[28] Quine represents a latter-day Boyle on this issue, for he says in a passage that could be a summary of Boyle's project:

> Each disposition, in my view, is a physical state or mechanism. A name for a specific disposition, e.g. solubility in water, deserves its place in the vocabulary of scientific theory as a name of a particular state or mechanism. In some cases, as in the case nowadays of solubility in water, we understand the physical details and are able to set them forth explicitly in terms of the arrangement and interaction of small bodies. Such a formulation, once achieved, can thenceforward even take the place of the old disposition term, or stand as its new definition.[29]

Both the deductive-nomological and the hypothetico-structural models are problematic, however. The DN model has well-known flaws and the HS model makes the assumption that a mechanistic explanation is always available. An alternative would be to allow the direct explanatory power of disposition ascriptions. Boyle would claim that this was unscientific but it is not unscientific according to more recent interpretations. Thompson has emphasized the importance of disposition concepts in modern physics.[30] He has said:

> position and velocity should now be related not to spatial properties or actual shapes, but to propensities. . . . Quantum mechanics uses the 'propensity' type of disposition, as this displays its effects probabilistically.

[25] 'About the Excellency and Grounds of the Mechanical Hypothesis', 147.
[26] 'The Origin and Forms of Qualities', 25. [27] Ibid. [28] Ibid. 26.
[29] W. V. O. Quine, *Roots of Reference* (La Salle, Ill., 1974), 11.
[30] I. J. Thompson, 'Real Dispositions in the Physical World', *British Journal for the Philosophy of Science*, 39 (1988).

> . . . If we then ask what must the world be like in order that quantum mechanics describes it correctly, we arrive at the existence of real propensities . . . . the notion [of a disposition] is likely to be fundamental to a realistic and non-paradoxical account of quantum physics . . . . it is thus important to resist certain interpretations of physics and of the physical world that render dispositions impossible. . . . In quantum field theory (a more complete form of quantum physics), even the *existence* of objects is a dispositional property that may or may not be manifested, as, for example, pairs of particles and anti-particles may or may not be formed.[31]

Thompson is not alone in this interpretation of physics,[32] which if correct means that physical theory requires a creditable explanatory role for dispositional ascriptions. Appeal to dispositions becomes the ultimate appeal in explanation.

1.4 *Causally Efficacious Properties versus Conditional Analyses*

What is it in virtue of which dispositions are alleged to have a role in explanation? Restricting this question to the case of concrete dispositions, initially, I will be arguing that appeal is made to them in explanation on the basis of two features. The explanatory value of appeal to dispositions typically resides in them being *causally efficacious* and being *properties*. There is good reason to regard these two features as coming together in a package (Sect. 6.2, below) but until I have justified this claim I will treat them separately.

The justification for regarding dispositions as causally efficacious is as follows. If type-identical stimuli are applied to two objects and one reacts differently from the other, then the difference in reaction must be accounted for in terms of some difference between the objects and this is a difference that has a causal effect on the reaction. For instance, two white cubes may be placed in two identical quantities of a liquid which are subject to the same background conditions. One of the white cubes dissolves, the other does not. The most obvious explanation for this difference in beha-

[31] 'Real Dispositions in the Physical World', 76–7.

[32] See e.g. Popper, 'The Propensity Interpretation of the Calculus of Probability'. H. Robinson, *Matter and Sense* (Cambridge, 1982), ch. 7, makes much the same point about the reliance of modern physics on disposition concepts but, as we will see in Ch. 2, he thinks this a reason to say that modern accounts of matter are incoherent.

viour is that one of the cubes was soluble in these conditions and the other was not. For there to be explanatory value in this disposition ascription it must mean more than that such dissolvings occur in those circumstances. Rather, the ascription of solubility would have to be taken as an ascription of something that is causally efficacious of such behaviour in such conditions, namely, a *property* of the object. This is to contradict the Ryle–Wittgenstein view.

The point of contention is whether dispositions are really properties or whether they are reducible to a different class of entities such as events. Dispositions are appealed to as if they were properties but this is a basic error according to the Ryle–Wittgenstein view. Ryle claims that the meaning of a disposition ascription can be explicated without imputing specifically dispositional properties. Ryle says:

> A statement ascribing a dispositional property to a thing has much, though not everything in common with a statement subsuming the thing under a law. To possess a dispositional property is not to be in a particular state, or to undergo a particular change; it is to be bound or liable to be in a particular state, or to undergo a particular change, when a particular condition is realised.[33]

Despite Ryle's reference to dispositional *properties*, it is clear that he regards the construal of dispositions as properties, or as he says 'states', as a misunderstanding of the logic of disposition ascription. Instead, Ryle opts for an analysis or reduction of disposition ascriptions in terms of actual and possible events specified in the antecedent and consequent clauses of hypothetical 'if . . . , then . . .' statements. This is made more explicit when he says:

> [T]he vogue of the para-mechanical legend has led many people to ignore the ways in which these concepts actually behave and to construe them instead as items in the descriptions of occult causes and effects. Sentences embodying these dispositional words have been interpreted as being categorical reports of particular but unwitnessable matters of fact instead of being testable, open hypothetical and what I shall call 'semi-hypothetical' statements.[34]

[33] *The Concept of Mind*, 41. Compare Wittgenstein, *Blue and Brown Books* (Oxford, 1958), 101: 'to this "state" there does not correspond a particular sense experience which lasts while the state lasts. Instead of that, the defining criterion for something being in this state consists in certain *tests*.'

[34] *The Concept of Mind*, 117.

Ryle analyses disposition ascriptions into complexes of possible events specified in conditionals. Thus an ascription of solubility would mean, on this analysis, 'if x is put in liquid, x dissolves'. Similarly with any other disposition ascription: 'x is fragile' means 'if x is knocked or dropped, x breaks', 'x is live' means 'if x is touched by a conductor, then electrical current flows from x to the conductor'.[35] In general, for disposition D, test condition F and confirming reaction G:

$$[\mathrm{Df}_R] \quad \forall x\ ((\mathrm{D}x) \leftrightarrow (\mathrm{F}x \rightarrow \mathrm{G}x)),$$

where the conditional '$\rightarrow$' is understood as a subjunctive. We can call this the conditional reduction or conditional analysis of dispositions or, for reasons that will become clear, simply *the empiricist view*.

Geach raised a number of questions about Ryle's view of dispositions.[36] It appears that responses can be found to these questions but I will show, in Sect. 2.2, that they are answered at the cost of accepting an implausible theory of dispositions.

First, Geach says that such an analysis of dispositions means that they are no longer causes of the disposition manifestations; in fact the very term 'disposition-manifestation' must, if Ryle is correct, be a philosophical confusion. If disposition ascriptions are reports of actual and possible events, consequent upon other (test) events, then they merely report those events—they state that they are or could be so—without offering any indication of *why* they are so. Thus what explanatory role is left for dispositions? It appears none. However, the Rylean can accept, without any immediate cost, that disposition ascriptions are not causally explanatory. Given that an alternative and satisfactory explanation is available, for instance, the presence of some underlying state of the object, then there is no reason to require explanation from a disposition ascription. If their analysis of disposition ascriptions is correct, therefore, the Rylean rejects the attempt to explain by reference to dispositions.

The second question asked by Geach is related to this. Geach states that when two particulars differ in behaviour we look for some actual, not merely hypothetical, difference between them.

[35] This example from C. B. Martin, 'Dispositions and Conditionals', *Philosophical Quarterly*, 44 (1994) will be discussed at length in Chs. 3 and 4.

[36] P. T. Geach, *Mental Acts* (London, 1957), 4–7.

Certainly there is a hypothetical difference between them if Ryle is right, but not necessarily *merely* a hypothetical difference. There could be an important difference between them in structural properties—perhaps molecular structural properties—that accounts for the difference in behaviour. Ryle's point is that what is alluded to in a disposition ascription is the actual or possible behaviour, not such underlying states, for these may be non-dispositional.

Finally, Geach wonders what the utility of the conditional analysis is. Certainly we have reduced a problematic mode of discourse—disposition ascription—but we have reduced it to a mode of discourse which is equally problematic: hypotheticals. There is as yet, we are told, no logic of hypotheticals, so nothing is gained in resorting to them. Certainly hypotheticals are a problem area, notwithstanding the significant contributions to their analysis since Geach wrote,[37] but it seems that Ryle could at least be credited with putting the problems in their correct places and order of priority. He has shown that a full theory of disposition ascription would have to rely upon a theory of hypotheticals, but not vice versa. The option is left open, however, for an empiricist theory of the truth of hypotheticals that makes no appeals to possibilia.

Given these answers to Geach, should we accept Ryle's view that dispositions are not causally efficacious properties and not explanatory? I will argue that we should not, but this will involve a deeper look at Ryle's position as well as the role of conditionals in the meaning of disposition ascription and the empiricist assumptions that underlie such treatments. The intuitive motivation for taking dispositions to be causally efficacious remains strong despite the Rylean challenge. I will set this challenge aside for the moment, however. This is not to avoid problems because I do so to bring up still further problems.

1.5 *The Metaphysical Question*

Let us entertain the assumption that there are dispositional properties possessed by things. Let us also accept that some properties,

[37] R. Stalnaker, 'A Theory of Conditionals' (1968), repr. in F. Jackson (ed.), *Conditionals* (Oxford, 1991); D. K. Lewis, *Counterfactuals* (Oxford, 1973); and E. J. Lowe, 'The Truth about Counterfactuals', *Philosophical Quarterly*, 45 (1995).

contrasted with the dispositional, are categorical properties. How would these two types of property be related? Would they stand completely apart or would there be some degree of interaction between them? If so, what sort of interaction? Which of the properties would be the causes of the events we ordinarily call disposition manifestations? If it is the dispositional, then the categorical become causally impotent. If it is the categorical, then the dispositional are causally impotent and the term 'disposition manifestation' is a misnomer. Could it be, then, that the dispositional and categorical properties conspire together to have effects jointly? If so, then we would have to say that neither type of property was causally sufficient for such events.

This collection of problems should sound familiar for there are parallels in other areas, most notoriously in the philosophy of mind. The allowance of two distinct types of property brings the demand for some account of the relation in which they stand. We will have to commit ourselves to an ontology. Some of the positions available are:

A. Property Dualism: *The ontological thesis that dispositional properties are a fundamentally different type of property from categorical properties.*

There are separate dispositional and categorical properties inhabiting the world, thus we have a dualism of types of properties which is the parallel to Cartesian dualism in the philosophy of mind. Given that few who have written on the subject are sensitive to the issue, categorization of previous views is not easy. However, some who might be classed as property dualists are Prior, Pargetter, and Jackson; Franklin; Weissman; Thompson; and Broad.[38]

B. Property Monism: *The ontological thesis that there is only one fundamental type of property.*

Any distinction drawn between types of properties cannot, therefore, be an ontological division in reality similar to that alleged to divide mental and physical properties. As we know in

[38] E. Prior, R. Pargetter, and F. Jackson, 'Three Theses about Dispositions', *American Philosophical Quarterly*, 19 (1982); J. Franklin, 'Are Dispositions Reducible to Categorical Properties?', *Philosophical Quarterly*, 38 (1988); D. Weissman, *Dispositional Properties* (Carbondale, Ill., 1965); I. J. Thompson, 'Real Dispositions in the Physical World'; C. D. Broad, *The Mind and its Place in Nature* (London, 1925), ch. 10.

the case of philosophy of mind, however, there can be more than one form of monism. Thus, there is a further choice for the monist to make.

B(i). Categorical Monism: *The ontological thesis that there is only one fundamental type of property. All properties are categorical properties; 'dispositional' properties do not exist.*

Proponents of this view, though not explicitly, are arguably Quine, Mackie, and Armstrong.[39]

B(ii). Dispositional Monism: *The ontological thesis that there is only one fundamental type of property. All properties are dispositional properties; 'categorical' properties do not exist.*

Arguable proponents: Popper; Mellor, and Shoemaker.[40]

There are, admittedly, certain disanalogies between the cases of mind and dispositions. Psychophysical interaction evidently occurs but a dualist has the problem of explaining how something that has no physical location or extension can interact with something located and extended in space. In the case of dispositions, this particular problem of interaction does not apply. Locke gave one example of such an interaction when he said: 'Pound an Almond, and the clear white *Colour* will be altered into a dirty one, and the sweet *Taste* into an oily one. What real Alteration can the beating of the Pestle make in any Body, but an Alteration of the *Texture* of it?'[41] If texture is taken as a categorical property and colour as a distinct dispositional property, it is plausible also to class both as physical properties and so there is no immediate problem parallel to the one of interaction between physical properties and unlocated and unextended properties.

A dualist ontology for dispositions is, therefore, not infected with one of the problems of psychophysical dualism but it is infected with another. This concerns the problem of causal efficacy

[39] W. V. O. Quine, *Roots of Reference*; J. L. Mackie, *Truth, Probability and Paradox* (Oxford, 1973), ch. 4 and 'Dispositions, Grounds and Causes', *Synthese*, 34 (1977); D. M. Armstrong, *A Materialist Theory of the Mind* (London, 1968), 85–9 and *Belief, Truth and Knowledge*, ch. 2.

[40] K. R. Popper, 'The Propensity Interpretation of the Calculus of Probability', and *The Logic of Scientific Discovery* (London, 1959), app. 10; D. H. Mellor, 'In Defense of Dispositions'; S. Shoemaker, 'Causality and Properties', in *Identity, Cause and Mind* (Cambridge, 1980).

[41] J. Locke, *An Essay Concerning Human Understanding*, 2. 8. 20.

and the possible overdetermination of effects. As Peacocke has suggested, we may believe that there is a wholly physical cause of someone raising their hand, consisting in the minute intricacies of what goes on in their brain.[42] But if so, what room is there for the *intention* to raise their hand to be a cause of the same event? The problem goes both ways. If the volition is the cause of an action, then it seems that any causal role for brain states, on a dualistic ontology, leads to a problem of overdetermination. The analogous problem in the case of dispositions is suggested near the start of this section. If a physical theory provides a complete causal explanation of events in terms solely of categorical properties, then the causal efficacy of dispositions on the same events can be allowed only at the cost of the overdetermination of those events.

This sort of problem forces us to consider the metaphysical question for dispositions. It is hoped that by the end of the present study a convincing ontology for dispositions will have been found and that it will be an ontology that makes room for dispositions and their causal potency in the physical and mental realms without contravening any common-sense or scientific commitments.

1.6 *Contrasts with Dispositions*[43]

It is not just the notion of the dispositional that stands in need of clarification but with it, also, the categorical. What does the non-dispositional consist in? There are at least two predominant opinions. For some, the notion of a disposition is contrasted with the notion of an occurrence; for others, it is contrasted with that of a categorical property. Each contrast is as reasonable to make as the other, though the clarity and coherence of each has been challenged. In evaluating the contrast between the dispositional and the non-dispositional, however, we need to be careful to keep the conceptual and ontological questions separate. One possible position has it, for instance, that there is a conceptual distinction between the dispositional and the categorical though there is no corresponding ontological division between properties. I will,

[42] C. Peacocke, *Holistic Explanation* (Oxford, 1979), 134ff.

[43] I am indebted to E. J. Lowe for raising in correspondence the concerns I give voice to in this section.

therefore, be treating the ontological and conceptual questions separately.[44]

First there is the disposition–occurrence distinction, which is the distinction that Ryle makes, where occurrences are sometimes referred to as episodes.[45] For Ryle, the contrast is stated as follows. A disposition statement is not a report of a current episode but a statement that such episodes *tend* to occur. Such episodes, occurrences, or events have a logical priority over dispositions for, we are told, 'unless statements like the first [reports of events] were sometimes true, statements like the second [disposition ascriptions] could not be true'.[46] Thus there are no true disposition ascriptions unless there are events (occurrences, episodes).

The problem with this empiricist approach is that realist intuitions seem to have eminent credibility. Perhaps not in the case of being a smoker, but certainly in the case of an object being fragile, it seems plausible that a disposition could be possessed though no manifestation events occur. Dispositions seem to be things which endure between events and this is a starting point for the realist attack on the conditional analysis (Chapter 3, below).

Second there is the contrast between a dispositional and a categorical property.[47] There is a lot of work to be done to make this distinction lucid. In the first place, it can plausibly be asserted that dispositions are categorical in that their ascriptions are true of their possessors unconditionally and 'categorical' just means 'unconditional'. Certainly there is an element of conditionality involved in dispositions, but what is conditional and potential is the *manifestation* of the disposition, not the disposition itself. It is the breaking of the glass that is potential; the fragility it possesses is possessed actually. Thus for someone like C. B. Martin, who has frequently stated that dispositions are 'flatly existent',[48] the dispositional is as

[44] There is one connection between the two issues however, namely that a property dualist must support the view that the conceptual distinction between dispositions and non-dispositions is clear, for there cannot be an ontological division if there is not even a conceptual one.

[45] It is the distinction also emphasized by E. J. Lowe, 'Laws, Dispositions and Sortal Logic', *American Philosophical Quarterly*, 19 (1982).

[46] *The Concept of Mind*, 117.

[47] e.g., Armstrong, *A Materialist Theory of the Mind*, 85–8.

[48] From C. B. Martin, 'The Need for Ontology: Some Choices', *Philosophy*, 68 (1993), 520. See also C. B. Martin, 'Anti-Realism and the World's Undoing', *Pacific Philosophical Quarterly*, 65 (1984), 'Power for Realists', in J. Heil (ed.), *Cause, Mind and Reality* (Dordrecht, 1993) and 'Dispositions and Conditionals'.

categorical as anything else. This means that anyone who seeks to defend the dispositional–categorical distinction has either to construe dispositions as mere possibilia, which I think would be a mistake, or to construe the categorical in different terms from those I have just outlined.[49]

One such way that the categorical has been construed is as a structural, or even 'geometrical-structural', property.[50] Again, inadequate description of structural properties is given. Montuschi has gone so far as to say that 'structure' is an inherently dispositional notion for nothing is gained for explanation by appealing to structures unless the structure is itself subject to causal necessity.[51] Clearly a lot of work has to be done and it had better be done first. The ontological issues will have to be set aside until it has been settled whether the dispositional and categorical mean different things.

1.7 *Strategy*

I will be tackling these problems in what I think is the most logical order. I start with an examination of the most threatening form of reduction for dispositions, the reduction of dispositions to conditionals, which has the ontological consequence that there are no dispositions *qua* properties. The conclusion will be anti-reductionist and will assert that there is something more in the meaning of a disposition ascription than is included in a conditional, or set of conditionals, relating possible events.

Having made a case for the existence of dispositions I then consider the contrasts that are drawn between dispositions and occurrences, structures and categorical properties. I will attempt to state a reasonably clear-cut criterion with which we can distinguish dispositional properties from other things.

Once the conceptual distinction has been established, work can be begun on the ontological question. In Chapter 5, I examine the

[49] This problem is not considered at all in one defence of the distinction: E. Prior, 'The Dispositional/Categorical Distinction', *Analysis*, 42 (1982).

[50] D. Weissman, 'Dispositions as Geometrical-Structural Properties', *Review of Metaphysics*, 32 (1978).

[51] E. Montuschi, 'From Effects to Causes: the Role of "Structure" in Scientific Explanation', *Conceptus*, 25 (1991).

ontology of property dualism. I will put forward the arguments for the *distinctness thesis* and show that they have a degree of force. I will go on to sketch a dualist ontology but show that there are potential problems. I will show that we have a plain choice between rejecting the distinctness thesis and accepting that certain properties are causally impotent.

Before allowing this to lead us to a property monism, by way of *reductio*, there needs to be some justification for the claim that dispositions are to be classed as causally potent. There are a number of very serious empirical and philosophical arguments contrary to this view which, if they are correct, mean that the causal impotence of dispositions could be conceded at no price. The most famous such argument is Molière's joke at the expense of the philosophers: the *virtus dormitiva* argument. I will argue that this does not undermine the claim that dispositions can be understood as causally efficacious.

In Chapter 7, property monism is presented. There will be consideration of what is required for the defeat of property dualism and an argument for identity of dispositional and categorical property tokens will be advanced which is an adaptation from a similar argument in the philosophy of mind. This is an argument from identity of causal role, and some objections to it will be scrutinized.

Next I move on to consider what form of monism we should support. I consider two alternatives: categorical monism and dispositional monism. I will examine the various possible justifications of these two positions, which must be motivated either by reductionism or eliminativism. I will support neither categorical nor dispositional monism so the qualitative ontological question, concerning the dispositional or categorical nature of properties, remains. My solution to this situation will lie in a correctly described functionalist theory of dispositions where the dispositional and non-dispositional are understood as two distinct ways of denoting the very same instantiations of properties.

Having established a dispositional–categorical distinction and described an ontology for dispositions, finally I show how dispositions so construed can replace the problematic notion of a law of nature as the fundamental building block of science. Hence dispositions will have been made respectable philosophical entities to which science can appeal.

First, however, I will introduce a number of other problems that arise from appeals that are made to dispositions in various philosophical disputes. These disputes are in the areas of philosophy of mind and philosophy of matter.

2

Dispositions in Mind and Matter

2.1 *Two Appeals to Dispositions*

As a prelude to the theory of dispositions that will be developed over the following eight chapters, I introduce two cases where appeal has been made to dispositions in philosophical disputes: in the philosophy of mind and the philosophy of matter. In the area of mind, understanding certain mental phenomena as dispositional seemed to offer an attractive alternative to Cartesianism. The alternative was attractive insofar as it made the mental publicly observable, in contrast to the privacy of the Cartesian ego. Problems stand in the way of a dispositional analysis, however. First, there is the issue of whether all mental phenomena can be understood completely as dispositions. A controversial question concerns whether there is a qualitative aspect of experience which is necessarily resistant to a dispositional analysis. The second problem concerns finding the right account of (mental) dispositions. Concerning the philosophy of matter, there are questions about the primary–secondary qualities distinction, the basis of fundamental laws, and, what I shall consider here, the nature of matter where dispositions play a crucial role. I will show in this chapter that, although appeal to dispositions may be the right first step to take in these disputes, if such appeal is to bring benefits, it is essential that it be backed up with a credible account of what such dispositions are supposed to be.

This chapter consists in two case studies where an initial appeal to dispositions is not contested but where such appeal is not backed with the correct theory of dispositions and ultimately unsatisfactory results emerge. The two cases involve appeals to dispositions understood in the empiricist way. I will argue that a realist account of dispositions would be better. I will be saying only a minimum about the realist theory at this stage, though. It is not to be thought that this discussion constitutes the argument for a realist theory of dispositions. A satisfactory argument for realism can emerge only at greater length.

2.2 *Ryle on the Dispositional Analysis of Mind*

One appeal to dispositions is in the philosophy of mind where, some have argued, to ascribe certain classes of mental predicates is to say only that certain dispositions are held by the subject of ascription. It is not to say that the subject of ascription is undergoing a private mental occurrence. To say that *x* understands algebra, for instance, is just to say that *x* is disposed to find correct solutions to algebraic equations if they are presented to *x*. A dispositional analysis of mental ascriptions has the putative advantage of making the mind publicly observable, rather than a private inner theatre of consciousness accessible only first-personally, because whether such dispositions are possessed can be tested when certain situations occur.

A dispositional analysis of *certain* mental phenomena is, I think, a correct way of understanding the mind. However, for an advance to have been made it is also necessary that a plausible account be given of what the possession of a disposition consists in. The conscious target of Ryle's new philosophy of mind was the Cartesian myth of the ghost in the machine which he sometimes calls 'the vogue of the para-mechanical legend'.[1] However, Ryle's account of dispositions was one that had the flaws of being severely reductive, empiricist, and anti-realist. This account is one that I take to be problematic. Given the attraction of a dispositional analysis in this area, the problems in Ryle should motivate a search for a better theory.

According to Ryle, there are no dispositions *qua* properties. There are only events that are witnessed to stand in certain contingent relations. An episode, occurrence, event, or state may be observable but a disposition is not. To have a belief is to have a disposition, which is neither an occurrence nor a state, hence it is not observable. Although unobservable, the ascriptions of disposition concepts are, however, testable:

[1] *The Concept of Mind*, 117.

Sentences embodying these dispositional words have been interpreted as being categorical reports of particular but unwitnessable matters of fact instead of being testable, open hypothetical and what I shall call 'semi-hypothetical' statements. The old error of treating the term 'Force' as denoting an occult force-exerting agency has been given up in the physical sciences, but its relatives survive in many theories of mind.[2]

Ryle's account is one that can be considered a paradigm of an empiricist treatment of dispositions. Dispositions are not reports of 'limbo-facts'. Limbo-facts would be ascriptions of potentialities and 'Potentialities, it can be truistically said, are nothing actual.'[3] Instead, Ryle gives an account where the truth of a disposition ascription consists in the truth of a hypothetical or subjunctive 'if . . . , then . . .' proposition because the conditional is the *analysans* of the disposition ascription. To say that something is brittle, according to Ryle, means nothing more nor less than that if that thing is ever struck or strained, then it would fly apart. Mental disposition terms are treated in exactly the same way as non-mental terms, analysed in conditionals, with an admission that in the case of the mental the *analysans* may contain greater complexity. This complexity consists in the fact that 'Dispositional words like "know", "believe", "aspire", "clever" and "humorous" are determinable, rather than determinate, dispositional words. They signify abilities, tendencies or pronenesses to do, not things of one unique kind, but things of lots of different kinds.'[4]

We may still think that something is left unexplained. Ryle asserts that 'The brittleness of glass does not consist in the fact that it is at a given moment actually being shivered. It may be brittle without ever being shivered.'[5] Things can rightly be said to have dispositions even though there is no current manifestation of that disposition; indeed, a particular may have a disposition that it has never manifested as in the case of a vase that has been fragile throughout its existence. The truth of the ascription of fragility consists in no state of affairs: it consists in nothing more than that a certain hypothetical has been true for the duration of that time, even if it is an untested hypothetical.

But what, we may now wonder, makes such a conditional true? One tempting answer to this question is that it is some property of the subject of ascription that makes the ascription true by endowing

[2] Ibid. 117. [3] Ibid. 119. [4] Ibid. 118. [5] Ibid. 43.

its possessor with the right causal powers. This is indeed Armstrong's answer to the question[6] and an answer that I suggested could be given in response to one of Geach's criticisms (Sect. 1.4, above). It is an answer that is not an option for Ryle, however, given his empiricist commitments.

Ryle has asserted that to have a disposition is not to be in a particular state. Indeed there would be little point in identifying such a property or state, for Ryle, given that he has the empiricist suspicion of causal connections between properties as anything more than regular associations of events. This is implicit when he argues that the explanations we give are actually appeals to further contingent regularities, not to properties:

> It is sometimes urged that if we discover a law, which enables us to infer from diseases of certain sorts to the existence of bacteria of certain sorts, then we have discovered a new existence, namely a causal connection between such bacteria and such diseases; and that consequently we now know, what we did not know before, that there exist not only diseased persons and bacteria, but also an invisible and intangible bond between them. . . . Now there is no objection to employing the familiar idiom of 'causal connection'. Bacteriologists do discover causal connections between bacteria and diseases, since this is only another way of saying that they do establish laws and so provide themselves with inference-tickets which enable them to infer from diseases to bacteria, explain diseases by assertions about bacteria, prevent and cure diseases by eliminating bacteria and so forth. But to speak as if the discovery of a law were the finding of a third, unobservable existence is simply to fall back into the old habit of construing open hypothetical statements as singular categorical statements.[7]

Dispositions, like laws, license inferences from one state of affairs or occurrence to another. Empiricists are at liberty, though not compelled, to take the truths of their ascriptions as basic facts about the world, just as they are at liberty to take the laws of nature as basic facts that can themselves have no further explanation. Armstrong's 'truth-maker argument', that every true contingent proposition must have something that makes it true,[8] need not carry weight in Ryle's empiricism and thus would not compel him

[6] *A Materialist Theory of the Mind*, 85–8. [7] *The Concept of Mind*, 122.

[8] D. M. Armstrong, 'C. B. Martin, Counterfactuals, Causality, and Conditionals', in J. Heil (ed.), *Cause, Mind and Reality: Essays Honoring C. B. Martin* (Dordrecht, 1989), 9.

to accept entities such as 'categorical bases' of dispositions. Ryle's position appears to be that affirmation of anything more than conditionals is going beyond what is empirically knowable and flirting with unobservable entities and occult potentialities.

What is wrong with Ryle's assumptions? A problem comes when we consider the difference between two particulars that differ in the disposition ascriptions that we think can truly be made of them. We want some actual, not merely hypothetical, difference between them. On Ryle's conditional analysis, there would be no actual difference between two untested particulars that differed in possession of a disposition because Ryle does not accept the possible realist answer that is available to this problem. A solely conditional difference is not difference enough if there is no actual property or state that is making the conditional true.

Disposition ascriptions can be tensed, as Ryle allows,[9] meaning that something may have been fragile one day, then non-fragile the following day, and then fragile again the next. What if such an object is not struck or stressed in any suitable way throughout these times? How is the change of disposition to be understood? The Rylean explanation will merely be that on the first day a conditional was true of the object which was not true on the second day but was true again on the third. Armstrong's account will accept this but offer the further explanation that the change in the truth-value of the conditional was a result of a change in the properties of the object during the days concerned. This is a difference in what is actual, on the relevant days, that explains the differences in possibilities and this might reasonably be thought to be an improvement on an account that describes the difference in dispositions only in terms of differences in possibilities.

However, given Ryle's empiricist objection to unobservable causal connections it might be thought that this version of the truthmaker argument will not sway the convinced Rylean. In answer to this, in the next chapter I will examine the claim that it is possible to provide a conditional analysis for all disposition terms and I will argue that there are strong realist motivations against the view. This tempts us to lean towards something like Armstrong's account where a disposition has a 'categorical base'. Dispositions,

[9] *The Concept of Mind*, 122.

understood as real properties, will be made legitimate entities to which appeal in philosophy of mind can be made.

2.3 *The Ontological Status of Matter*

A second philosophical problem that has the notion of a disposition at its centre is the question of the existence of matter. Howard Robinson argues for idealism by attacking the coherence of the existence of matter.[10] The attack on matter rests on the claim that anything physical can be characterized dispositionally only. All that we can say exist categorically are minds with mental states as their objects. The same problem of matter has been touched upon by Blackburn where, unlike Robinson, the self-sufficiency of dispositions is endorsed.[11] It is, indeed, in Robinson's understanding of dispositions that his argument falters. It may well be correct to say that matter is understood dispositionally but this needs to be supported by the correct account of what dispositions are. If we treat dispositions correctly, we can save the material world from Robinson's attack.

Robinson begins by looking at the philosophical attempts to characterize matter: to find a simple, unconditional, and categorical account of what it is to be a physical thing. His initial concern is restricted to conceptual analysis. Descartes said that extension was the essence of material bodies: a body is a geometrically definable space.[12] This is obviously an incomplete characterization. The idea of extension provides only the idea of a bare geometrical figure which may just be a volume of empty space. What needs to be added to Descartes' idea of extension is that the extended space be an occupied one. Locke attempted to make such an addition by including solidity in his list of primary qualities, the others being extension, figure, motion or rest, and number.[13] Only by the addition of solidity to such 'geometrical' qualities is the area of space made an occupied one.

Robinson's critique of this analysis is not new. It can be found in Hume's *Treatise of Human Nature*. Hume points out that our ideas

[10] *Matter and Sense*, ch. 7.
[11] S. Blackburn, 'Filling in Space', *Analysis*, 50 (1990), 62–5.
[12] R. Descartes, *Meditations on First Philosophy* (1641), second meditation.
[13] J. Locke, *An Essay Concerning Human Understanding* (1690), 2. 8. 9.

reveal only a property of impenetrability and that impenetrability is a dispositional property:

In order to form an idea of solidity, we must conceive two bodies pressing on each other without penetration; and 'tis impossible to arrive at this idea, when we confine ourselves to one object, much more without conceiving any.[14]

Without any independent idea of solidity we have no coherent account of matter, for no other supposed quality can fill the role solidity is supposed to play:

Now what idea have we of these bodies? The ideas of colours, sounds, and other secondary qualities are excluded.[15] The idea of motion depends on that of extension, and the idea of extension on that of solidity. 'Tis impossible, therefore, that the idea of solidity can depend on either of them. For that wou'd be to run in a circle and make one idea depend on another, while at the same time the latter depends on the former. Our modern philosophy, therefore, leaves us no just nor satisfactory idea of solidity; nor consequently of matter.[16]

We have arrived at a conception of material bodies as volumes of impenetrability. Impenetrability is a dispositional property only. It describes how an object will act in certain circumstances when it collides with another body of its own kind, not what it is in itself. The crucial premiss here is that the dispositional properties of a thing do not provide us with its real nature.

The argument then proceeds to a Berkeleyan conclusion. Matter is understood only in terms of dispositions to produce effects. The effects matter produces are the perceptible qualities which are accessible to our understanding. Such perceptions are the only things which are actual or categorical because, as Robinson established in preceding chapters, they are the only things that cannot be further reduced to dispositions or potentialities. Furthermore, such perceptible qualities are mind-dependent because they vary according to the perceptual apparatus of the perceiver. We arrive at the conclusion that 'the only categorical entities . . . are mental states with sensible qualities as their

[14] D. Hume, *A Treatise of Human Nature* (1739–40), 1. 4. 4.

[15] The reason for this being that '[t]he impressions, which enter by the sight and hearing, the smell and taste, are affirm'd by modern philosophy to be without any resembling objects; and consequently the idea of solidity, which is suppos'd to be real, can never be deriv'd from any of these senses.' Ibid.

[16] Ibid.

objects.'[17] Such mental states cannot be physical states as physicalists assert, for if they were physical states, then we would have to give them a dispositional account in terms of powers and no categorical base would have been provided at all. The postulation of any mind–brain identity thesis will, therefore, be self-defeating.

One could argue that solidity, the categorical grounding of matter, is not observable but is, rather, inferred from the observation of impenetrability. Robinson raises three objections to this. First, we would need to specify the relationship between the quality of solidity and the power of impenetrability and Robinson thinks that the connection is only a contingent one; it is nothing stronger than a Humean constant conjunction and, lacking any necessary connection, the two could possibly become separated. Without there being a necessary connection we can imagine solidity without impenetrability or impenetrability without solidity so the suggested analysis fails. Second, Robinson says that we are unjustified in inferring a categorical quality from a perceptible or dispositional quality, for such perceptible qualities are sense-receptor and mind-dependent thus, according to Robinson, lacking any objective status. Third, Robinson introduces the scientific account of matter to back up his conceptual analysis. He argues that any such solidity is only a feature of matter at the macroscopic level. If modern scientific accounts are to be believed, as materialists want us to believe, at the microscopic level there is no such thing as solidity.

Not surprisingly, Robinson construes the dispositions that play the crucial role in the argument along empiricist lines. This can be seen when Robinson spells out the stages of his regress argument, which proceeds as follows:[18]

(1) '*Every real (categorical) object* [entity, property] *must possess a determinate nature.*'

(2) '*The nature of any power P is given by what would constitute its actualization.*'

From which it follows that:

(3) '*If P is a real object it must be a power to a determinate actualization.*'

[17] *Matter and Sense*, 117. [18] Ibid. 114–15.

The regress becomes apparent in the following:

(4) '*If a power Q is the actualization of P, the determinacy of P will depend upon the determinacy of Q; that is, of Q's actualization.*'

(5) '*The list of effects constituting the determinate and complete nature of P will be finite only if the list contains (and thereby terminates at) an effect which is not a power.*'

and Robinson has argued for the case of matter that:

(R) *There can only be a list of powers for the nature of matter, so the nature of matter is indeterminate.*

A power is a power to alter another physical body; but that physical body is only a set of dispositions or powers to alter other physical bodies, which are powers acting on other physical bodies, and so on. Thus Robinson's claim: 'a power: P is the power to produce (alter) a power to produce (alter) a power to produce . . . Such a formula seems to tell us nothing about what is actually done.'[19]

Robinson has argued for two main theses. First, that powers exist only in virtue of a determinate actualization. Second, claiming the support of modern science, that physical entities are not categorical entities. Where, then, does Robinson go wrong in his understanding of dispositional properties that allows his regress argument to go through?

Robinson says that to ascribe a disposition is only 'to say how a thing will sometimes act' and 'not to say what it is'.[20] Such a view is akin to Ryle's empiricist analysis where the ascription of a disposition is nothing more than affirming the truth of a subjunctive conditional. Such a conditional analysis of dispositions gives the appearance that ascriptions of dispositions are only allusions to future possibilities and that dispositions are nothing in actuality. Robinson is clearly looking for the actualization of dispositions in the actuality of their manifestations. Thus, when matter is characterized only dispositionally, as impenetrable volumes of space, Robinson looks only to the manifestations of such impenetrability in the categorical properties that are mental states. We need not look for the actualization of a disposition solely in its manifestation, however, for we understand dispositions to be

[19] Ibid. 116. [20] Ibid. 108–9.

actual whenever they are ascribed. This is the realist alternative that I will be defending. If we were to treat dispositions as actual properties that play a causal role in their manifestations, then we can understand why dispositions are actual even when not currently manifested. Their actuality consists in the possession of a property which will play a certain causal role in certain circumstances.

This offers a way out of Robinson's regress. We can find the actuality of matter in the property or complex of properties and microscopic molecular structures which is causally efficacious of the observed macroscopic impenetrability when in contact with other matter. This is a property possessed whether or not there is any second object which comes into contact with a first and leads to the manifestation of impenetrability. What constitutes this basis of impenetrability will be understood only within a complex physical theory. It may be conceded that modern physics characterizes the microscopic facts of matter solely in terms of dispositional concepts: in terms of forces, fields, and energies. However, these in turn are to be understood as actual properties rather than just potentialities towards an actualization. At the microscopic level there may well be no impenetrability of the sort we consider at the macroscopic level but this need not mislead us. As Joske points out on this very issue, 'although we speak of elements, such as lead, as composed of molecules of lead, these molecules are not themselves leaden in the everyday sense of that word.'[21] The general principle that this suggests is that a property need not be present at the microscopic level in order to be present at the macroscopic level.

The attack on the existence of matter began from the assumption that dispositions are not real qualities of things and do not suffice to confer reality upon their subjects. Realism treats dispositions as real properties which inhere in their owners even when they are not manifested. The characterization of matter as an impenetrable volume of space can in this way be accepted as a legitimate and actual characterization of matter.

I have merely indicated the lines along which a realist theory of dispositions could be developed. A number of questions have been raised by these appeals to realism. In what sense and in what way are dispositions to be understood as real and actual? Do all dis-

[21] W. D. Joske, *Material Objects* (London, 1967), 15.

positions have a categorical base that is the truth-maker for any associated true conditionals? What is the relationship between such a base and the disposition itself? Are they distinct properties or states or are they two different ways of talking about the same properties? If the dispositional and the categorical just are two different ways of designating the same state, then in what way do they differ: what is being said that is different when a disposition is ascribed than when a categorical property is ascribed? In the next chapter I will start to answer these questions.

3

The Conditional Analysis

3.1 *Empiricism and Event Ontologies*

Let us take Hume as representative of the event ontologist and an event ontology as one possible expression of empiricism. Empiricism of this kind allows knowledge of nothing more than that which can be experienced or that which is a combination of experiences and suggests that it is events that are experienced rather than items such as substances and properties. Our notion of a property is typically said to be derived only via our experience of events and what we are capable of constructing out of such events.

The event ontologist, in general, has no inclination to accommodate dispositions *qua* properties into a description of reality. Events can, they allege, adequately account for all the purposes for which we would unnecessarily posit a dispositional property. Hume tells us that the distinction that is commonly drawn between a power and the exercise of it is without foundation, for all there is, in this instance, is events.[1] Hume makes his point in the specific case of dispositions when he says that: 'Power consists in the possibility or probability of any action, as discover'd by experience and the practice of the world.'[2] Giving a concrete example, according to the Humean account, to say that something is fragile is not to say that something has a hidden, 'occult' power. Rather, it is just to say that it is possible or probable that the thing will break. Similarly, our belief that sugar is soluble does not arise from an acquaintance with any secret faculties of the substance; it merely arises from our previous experience of sugar dissolving. If we are to be consistent empiricists, then all we ought to admit of is observation of such events. This is all that there is available to which an ascription of a power can refer. We have acquaintance

[1] *Treatise of Human Nature*, 1. 3. 14. [2] Ibid. 2. 1. 10.

with nothing else, hence the true reference of a disposition ascription is a set of events that are expected by us, given the evidence of previous events. The parallel with Hume's reductive account of causal relations is obvious. Our belief in real causal connections between events has no logical justification; it arises merely from custom and habit.

In this chapter I will be taking a close look at this view of dispositions and the conditional analysis of disposition ascriptions to which it leads. I will try to present it in as sympathetic a light as possible. Ultimately, however, this way of understanding dispositions will have to be rejected. The rejection will consist partly in a challenge to some of the underlying assumptions involved. The conclusion will be that the concept of a disposition is something that goes beyond complexes of observable events and, thus, that the conditional analysis fails to analyse or reduce disposition concepts. Eventually, however, it will be admitted that there is a connection between dispositions and conditionals, as the reductionists urge, but it will be shown that the relationship is not one of equivalence, contrary to their claims. This final issue will be the subject of Chapter 4. My aim here is restricted to the question of whether disposition ascriptions can be analysed wholly in terms of conditionals as empiricists have claimed.

3.2 *Dispositions and Conditionals*

Disposition ascriptions are categorical. Within the context of the previous literature on dispositions no statement could appear more confusing. The statement has the appearance of a contradiction because, ordinarily, categorical property terms are contrasted with dispositional and 'categorical' is used synonymously with 'non-dispositional'. I will explain what I mean when I say that disposition ascriptions are categorical and handle the question of where this leaves the putative dispositional–categorical distinction in the next chapter.

Disposition ascriptions are categorical in the sense that to say that something has a dispositional property is to say that something has a property *actually*. This is more than just the claim that disposition ascriptions are true for, as we saw with Ryle, it may be thought that a disposition ascription could be true without the claim that it is true in virtue of the possession of some property. Most typically, an ascription of a disposition is an ascription of a property to a particular *now* but really this is only the case where the disposition ascription is in the present tense and not all disposition ascriptions are present tense. We can make past-tense disposition ascriptions: 'the glass was fragile (before it was broken)'. We can make future-tense disposition ascriptions: 'the iron bar will be magnetic (after it has been magnetized)'. No matter whether the ascription is past, present, or future tense, however, the property ascribed is an actual property of the object. Hence a future-tense ascription would be of the form '*x* will be D, actually, at some future time *t*'.

If dispositions are to be construed as *possibilia* as some event ontologists would have it, that is, if disposition ascriptions are understood as merely indicative of possible events, then some explanation will be needed of why they are to be construed differently from their prima facie meaning. There is no immediate difference in logical form between the statement

[1] D*x*,

where 'D' is a disposition predicate, and an ascription of a non-disposition predicate 'C*x*'. Statements of the form [1] apparently have the same form as non-dispositional statements, so just as we could say '*x* is fragile', we could say '*x* is tall', '*x* is square', or '*x* is broken'. All these would have the single, monadic predication form in [1].

However, on the empiricist account, although [1] could be a correct representation of the ascription of predicates such as 'tall' and 'broken', it is not a correct representation of the logical form of a disposition ascription. Ryle, as we have seen, claims that '[t]here still survives the preposterous assumption that every true or false statement either asserts or denies that a mentioned object or set of objects possesses a specified attribute.'[3] Evidently, the statement '*x* is fragile' is true of some things and false of others but, according to Ryle, the truth or falsity of this ascription does not consist in *x* having some categorical property that we may try to ascribe with a statement of the form [1]. He explains, 'Dispositional statements are neither reports of observed

[3] *Concept of Mind*, 120.

or observable states of affairs, nor yet reports of unobserved or unobservable states of affairs. They narrate no incidents.'[4] Rather, disposition statements are 'lawlike' statements that are 'variable' or 'open'. For instance:

> To say that this lump of sugar is soluble is to say that it would dissolve, if submerged anywhere, at any time and in any parcel of water. To say that this sleeper knows French, is to say that if, for example, he is ever addressed in French, or shown any French newspaper, he responds pertinently in French, acts appropriately or translates it correctly into his own tongue.[5]

Ryle is not alone in making this type of claim for dispositions. Locke makes the point that for the ascription of a disposition, all we need know of is a certain reaction following in certain conditions; we do not need to know what makes the reaction occur in those conditions.[6] Dummett has more recently made a number of similar claims; for example, that 'the criterion for the possession of a dispositional or measurable property is that of giving a certain result on subjection to a particular test'[7] and 'A disposition of any kind is not a quality of which we can be directly aware . . . a disposition is something that is manifested in different circumstances.'[8] On such a view, certain theoretical explanations of why such events have occurred can be tolerated. What Ryle asserts however, is that a disposition ascription does not refer to a property of the object but to the fact that the object has actual or possible behaviour of a particular type.

Ryle and Dummett add something to the Humean account that obviously was required, namely, the claim that the events such as breaking, bending, and dissolving are dependent or conditional upon other events. These other events are the 'tests' and 'different circumstances' that Dummett mentioned. I drew attention to this feature of dispositions when I discussed fragility (Sect. 1.2). To say something is fragile is certainly to suggest the possibility of it breaking but it need never be the case that it actually breaks, for

[4] Ibid. 125. [5] Ibid. 123.

[6] *Essay Concerning Human Understanding*, 2. 21. 1. See M. R. Ayers, 'The Ideas of Power and Substance in Locke's Philosophy', in I. C. Tipton (ed.), *Locke on Human Understanding* (Oxford, 1977).

[7] M. Dummett, *Truth and Other Enigmas* (London, 1978), 150.

[8] M. Dummett, 'Realism', *Synthese*, 52 (1982), 111.

it may never be the case that it is dropped or knocked. In sum, the possibilities raised by a disposition ascription are consequent upon the realization of other, antecedent or stimulus, possibilities; hence Ryle's reduction of disposition statements into hypotheticals, or subjunctive 'if . . . , then . . .' statements. The truth of a disposition statement does not reside in the possession of a categorical property, as suggested in [1], but in the truth of a conditional of the form:

[2] If F*x*, then G*x*,

where the predicate 'F' is to be replaced by the antecedent test conditions for the disposition and 'G' by the appropriate manifestation. The predicate 'D' is true of *x* if and only if it is true that (if F*x*, then G*x*) and we dispense with the misleading use of 'occult' power-terms because the publicly observable complex of events named in the conditional completely exhaust the meaning of the disposition ascription. Examination of some typical examples demonstrates the plausibility of this position:

(a)	*x* is soluble	iff	if *x* is placed in liquid, *x* dissolves.
(b)	*x* is fragile	iff	if *x* is dropped or knocked, even lightly, *x* breaks.
(c)	*x* is magnetic	iff	if *x* is in the vicinity of a suitable metal *y*, *x* attracts *y*.
(d)	*x* believes P	iff	if *x* is asked whether P, *x* responds in the affirmative.
(e)	*x* is brave	iff	if *x* is aware of being in a frightening position, *x* does not flee.

(a) to (e) are all conditional analyses of the form [2]; that is, they are all proposed analyses where a single satisfaction of a single conditional, or the passing of a single test, is deemed sufficient for the truth of the disposition ascription.

This simplest of analyses may have to be admitted as a comparatively rare case, however, because in many cases of disposition ascriptions a single conditional will not suffice to analyse all the possible tests and manifestations for that disposition. The fact that many disposition terms cannot be analysed so simply does not immediately undermine the strategy for, where a disposition ascription would require analysis into more than one conditional, we could offer those conditionals as a conjunction or disjunction which analyses the disposition concept. Carnap considered electrical current and noted that there is more than one test for whether something is live. There is testing by 'measuring the heat produced in the conductor, or the deviation of a magnetic needle, or the quantity of silver separated out of a solution'.[9] A conditional reduction of such a term would, therefore, be of the form:

[3] Dx iff ((if F_1x, then G_1x) & (if F_2x, then G_2x) & (if F_3x, then G_3x) & . . .)

until all possible tests were accounted for.

The situation seems even more problematic than this for human dispositions such as bravery and love. If we follow Wittgenstein and take love to be a disposition, then it seems that love is most certainly a *variably manifested* disposition. It can manifest itself in different ways although there is difficulty in specifying exactly all the possible manifestations. We must allow, it appears, that for dispositions such as this there is a disjunction of possible manifestations; hence:

[4] Dx iff ((if F_1x, then G_1x) $\vee$ (if F_2x, then G_2x) $\vee$ (if F_3x, then G_3x) $\vee$ . . .).

There are two problems with such a definition, however. First, that it may never be possible to close the disjunction, suggesting that there is something in the meaning of the disposition term that cannot be captured by the conditional analysis; that is, something that cannot be analysed by a finite disjunction of conditionals. Second, questions are raised about how many conditionals in the disjunction it is necessary to satisfy for the disposition term to be correctly applied. The satisfaction of just one disjunct may be insufficient for the application of the disposition term; the satisfaction of them all may be unnecessary. In the case of love, for instance, buying someone flowers, if passing a florist, may be one of the disjuncts in the analysis of the ascription. This disjunct in clearly not sufficient for a case of love, for one may buy flowers for a colleague recuperating in hospital. It is not necessary for love because one may be aware of an allergy of one's beloved to flowers. Thus, although any attempt to analyse a disposition like love must

[9] R. Carnap, 'Testability and Meaning' I, *Philosophy of Science*, 3 (1936), 444–5.

be disjunctive, it cannot be disjunctive in the truth-functional sense where the disjunction is true if just one disjunct is true. The solution may require more than the satisfaction of a minimum subset of disjuncts, somehow specified, for the satisfaction of some disjunct-conditionals may be more important than the satisfaction of others; some, for instance, may be necessary conditions, others sufficient conditions. There are, therefore, numerous complications standing in the way of conditional analyses and in many instances an in-depth conceptual analysis would be required for the successful completion of the project.[10]

I put these problems aside for the present, though, because there are problems that can be raised for the simplest case of a conditional analysis that may spare us engaging in the complex questions of the conjunctive and disjunctive forms.

First, I consider further the consequence of an event ontology and what may be held along with it: a commitment to verificationism.

3.3 *Verificationism*

The empiricist treatment of disposition terms found expression in the verificationist philosophies associated with the logical positivists. However, verificationist understandings of disposition terms prove inadequate when it comes to certain notorious problem cases. The basic realist objection to the verificationist treatment of dispositions is that it cannot allow the intuitively plausible case of a true but unverified or unverifiable disposition ascription. This inadequacy of the verificationist treatment was exacerbated by the further commitment that some verificationists had to extensional logic.

How does verificationism arrive at a problematic account of dispositions? To understand this we need to consider two typical verificationist commitments.[11] First, the verification principle, which is a general claim about meaning: that a statement's meaning

[10] There is the further issue raised by J. C. D'Alessio, 'Dispositions, Reduction Sentences and Causal Conditionals', *Critica Revista Hispano Americana de Filosofia*, 14 (1967). Some dispositions may have necessary but not sufficient conditions, some sufficient but not necessary conditions, and some neither necessary nor sufficient conditions.

[11] O. Hanfling (ed.), *Essential Readings in Logical Positivism* (Oxford, 1981), 5.

consists in its method of verification. Second, the criterion of verifiability, which is a criterion for separating meaningful from meaningless statements on the basis of whether they are verifiable. Both of these commitments have a bearing on the verificationist treatment of dispositions and consequently they have had a big influence on philosophical accounts of dispositions in the early part of the century.

What was the significance of the criterion of verifiability to a verificationist account of dispositions? The verificationist had to consider the following. Logical positivism avowed confidence in the explanatory abilities of science and strove for the 'unity of science'; that is, a unification of all sciences with physical science as the most fundamental. Disposition predicates, however, abound in scientific discourse. We speak of things as 'malleable', 'elastic', 'flammable', 'showing reaction ψ under conditions φ', and so on. Our descriptions of the objects in the world seem to have this dispositional character; for we know of the qualities of things only through their observable manifestations and also we are interested in how things behave in a variety of different situations. Despite this abundance of disposition terms there is a problem. Being soluble is not the same as being dissolved. Dispositions are not the same things as their manifestations. If they are not equivalent to their manifestations, then what are dispositions? A disposition ascription looks prima facie to be the statement that something has a potentiality: that *x* can do something. Being soluble seems to mean, according to Locke, Hume, Ryle, and Dummett, something like 'the event of *x* dissolving is possible if *x* is put in liquid'. But how can we observe a possibility? Isn't there a strong case for saying that the possession of a disposition, construed pre-theoretically, is never observable; for all we can observe is its manifestation, never the thing itself?[12] The disposition term indicates something that can issue in observable events but is not the same thing as those observable events.

The verificationist's problem is this: it seems common sense that dispositions can be possessed over times when they are not manifested, and it seems plausible for at least some dispositions that

[12] I allow the possibility, as will become obvious later on, of a post-theoretical position that makes dispositions identical to something that is observable; thus in this special sense dispositions are observable also.

there is no time limit on how long these can be possessed unmanifested. How, then, can we verify the proposition that a particular possessed a disposition between times t_1 and t_2 when the disposition was never manifested between those times?

The logical positivists sought their verification of contingent singular propositions in 'observation sentences' or 'protocol statements', which were 'direct reports of experience'. But how can a potentiality be observed when it is not currently manifested? Ordinary categorical terms such as 'broken' or 'dissolved' have apparent simple empirical content as their applicability can be confirmed with a single observation. Single observations would certainly not suffice for the verification of disposition ascriptions. Unless some other way of verifying their truth could be found, in a number of observations, for instance, disposition ascriptions would be unverifiable in principle and so, on the criterion of verifiability, meaningless.

The second commitment that impacts on dispositions is the principle of verification. This brings a further problem for the verificationist: if the meaning of any proposition is given by its method of verification, then one conclusion that could be drawn is that all predicates take on a dispositional character. This is a result of the consequence that all propositions gain a hypothetical nature. For example, the proposition 'the cat is/was/will be on the mat at time *t*', will mean, for the verificationist, something approximating 'if a normal observer were to look at the mat at time *t*, then the cat will be on the mat'.[13] Such commitments are in sympathy with Popper's claim that 'all universals are dispositional':

> Thus 'broken', like 'dissolved', describes dispositions to behave in a certain regular or lawlike manner. Similarly, we say of a surface that it is red, or white, if it has the disposition to reflect red, or white, light, and consequently the disposition to look in daylight red, or white. In general, the dispositional character of any universal property will become clear if we consider what tests we should undertake if we are in doubt whether or not the property is present in some particular case.[14]

Thus to explain the significance of any predicate ascription, even those ordinarily construed as non-dispositional, is to describe its

[13] This consequence of verificationism is discussed by L. J. Russell, 'Communication and Verification', *Proceedings of the Aristotelian Society,* supp. vol. 12 (1934).

[14] *Logic of Scientific Discovery*, 424–5.

visual, tactile, audible, olfactory, and gustatory significance, which more strictly is nothing but its dispositions to affect our senses.

The line that has been pursued is one that is obviously going to lead to conditionals as supplying the truth-conditions of these statements. This sets us the task of making such truth-conditions for conditional statements explicit and this can only mean more problems for a theory of dispositions. What logical force do we want to give the conditionals we allude to in our disposition ascriptions?

3.4 *Material Implication*

There is a logic of material implication, a simple logic, and early verificationists used material implication in their attempts to define dispositions. First, let us distinguish two types of conditional which I will call the Philonian and the Diodorean. Philonian conditionals are truth-functional; that is, the truth-value of the conditional is determined for every case by the truth-values of the individual antecedent and consequent propositions.[15] This simplicity is a virtue of Philonian hypotheticals. As Peirce said, 'The utility . . . is that it puts us in possession of a rule . . . [namely] The hypothetical proposition may . . . be falsified by a single state of things, but only by one in which A [the antecedent] is true and B [the consequent] is false.'[16] The competing implication of Diodorus Cronos ruled that 'the connected (proposition) is true when it begins with true and neither could nor can end with false'[17] which is not a truth-functional conditional because its truth-value is not wholly determined by the truth-value of the constituent propositions, but also by what the particular constituent propositions are.[18]

[15] The Philonian conditional is named after Philo of Megara (*c.* 300 BC). Sextus Empiricus reports Philo's specification of the meaning of a conditional; see I. M. Bochenski, *A History of Formal Logic* (Notre Dame, 1961), 117.

[16] C. S. Pierce, 'On the Algebra of Logic: A Contribution to the Philosophy of Notation', in C. Hartshorne and P. Weiss (eds.), *Collected Papers of Charles Sanders Pierce*, iii (Cambridge, Mass., 1933), 218.

[17] Bochenski, *A History of Formal Logic*, 117–18.

[18] Two hypotheticals, both of which contain only true propositions, could, in Diodorean implication, have different truth-values. The propositions 'it is day' and 'I converse' may be both true but the implication 'if it is day, then I converse' false because it is possible that it be day and I be not conversing.

A Diodorean implication has, therefore, stronger conditions for its satisfaction than material implication.

Material implication was introduced into modern logic by Frege, without knowledge of Philo's ancient formulation, but having exactly the same truth-conditions. Frege took it as one of the two basic operators in his *Begriffsschrift.* The explicit claim is made that this connective is sufficient to represent causal relations. This he does when he says of the connective, having just outlined its Philonian truth-conditions: 'We can thus translate: "If something has the property *X*, then it also has the property *P*", or "Every *X* is a *P*", or "All *X*'s are *P*'s." *This is how causal connections are expressed.*'[19]

Causal connections are evidently appealed to in the case of the conditionals allegedly invoked in disposition ascriptions. Whatever the analysis of causation, the notion of something being soluble is a notion of something that is caused to dissolve when placed in liquid.[20]

3.5 *Carnap's Paradox*

Carnap showed that if we attempted to define dispositions interpreting the conditional as a material implication, then an apparent paradox would arise.[21] Where *t* is a bound time-variable, an initial analysis of disposition ascription may be:

$[\mathrm{D}_I]\quad \forall x\ \forall t\ (\mathrm{D}x \leftrightarrow (\mathrm{F}x,t \rightarrow \mathrm{G}x,t)),$

read as 'for any *x* and for any time *t*, *x* is D if and only if, if *x* is F at *t*, then *x* is G at *t*'. 'D*x*' is the disposition predication and for the example of solubility would be filled by '*x* is soluble', 'F*x*,*t*' would be '*x* is placed in water at *t*' and 'G*x*,*t*' would be '*x* dissolves at *t*'.

Carnap asked us to imagine that a wooden match *c*, that was completely burnt yesterday, was never placed in water at any time. Is it not the case, therefore, that *c* is soluble? This follows simply from D_I under its truth-functional interpretation. Because the

[19] G. Frege, *Begriffsschrift*, translated by T. W. Bynam as *Conceptual Notation* (Oxford, 1972), 134.

[20] The exact relationships involved between the concepts of disposition, base, cause, stimulus, and manifestation will be given a deeper consideration in Ch. 7.

[21] 'Testability and Meaning' I, 440.

antecedent of the material implication $Fx,t \rightarrow Gx,t$ is false, the material implication is itself true, by definition of material implication, which means that *c* satisfies the definition: the wooden match is soluble.

The adherent to the Fregean claim, that material implication is acceptable for the expression of a causal proposition, must also accept the following, however. Insolubility, taken to mean 'not soluble', would have to be defined as:

$$[D_2] \quad \forall x\, \forall t\, (\neg Dx \leftrightarrow (Fx,t \rightarrow \neg Gx,t)).$$

For the wooden match never placed in water, D_2 again has a false antecedent, thus a true material implication, and the definition D_2 is satisfied. Thus any object that is never immersed in water satisfies both D_1 and D_2 and is, therefore, both soluble and insoluble.

This contradiction can be taken to constitute a *reductio* of the position that dispositions can be defined in this way.[22] One conclusion that could be drawn is this: material implication is too 'weak' a conditional to be a correct interpretation of disposition ascriptions because dispositions cause their manifestations and a causal conditional is stronger than material implication. A causal conditional has counterfactual force, even if it is not always a counterfactual, and so is some version of a Diodorean conditional. A natural move, therefore, would be to abandon material implication.

Carnap did not introduce the paradox to promote the introduction of a non-truth-functional logic, though I will consider whether such a move would solve the problems of the conditional analysis (Sects. 3.6, 3.7, below). Rather, he attempted to offer a solution to the paradox within truth-functional confines. Carnap's response was to concede that dispositions cannot be defined truth-functionally but to argue that this did not matter as long as they could be *reduced* to a non-equivalent truth-functional sentence that was constructed out of observation statements. 'Testability and Meaning' is an attempt to integrate this project into the larger one of constructing a language for science based on observation. It

[22] Though the out-and-out extensionalist, who thinks that truth-functional logic suffices for all scientific and philosophical purposes, could claim that there is no contradiction in a particular being both 'soluble' and 'insoluble' as defined by D_1 and D_2. The main point is, therefore, that such a consequence would not suffice for ordinary usage of disposition terms.

attempts to show that 'all scientific terms could be introduced as disposition terms on the basis of observation terms either by explicit definition or by so-called reduction sentences, which constitute a kind of conditional definition.'[23]

The explicit definition is forgone and the 'reduction sentence' replaces it. The reduction sentence for a disposition term is given as:

[R] $\forall x\, \forall t\, (Q_1(x,t) \rightarrow (Q_3(x) \leftrightarrow Q_2(x,t)))$,

where Q_3 is filled by the disposition term. In the case of solubility the reduction sentence would be 'if any *x* is put into water at any time *t*, then, if *x* is soluble in water, *x* dissolves at the time *t*, and if *x* is not soluble in water, it does not.'[24]

A problem remains of which Carnap is fully aware. The problem with material implication is that whenever an antecedent is false—that is, whenever the test conditions are left unfulfilled—the material implication is true and the disposition can be ascribed. Reduction sentences do not fully eliminate this difficulty because there still needs to be some adjudication on disposition ascriptions

[23] R. Carnap, 'The Methodological Character of Theoretical Concepts', *Minnesota Studies in the Philosophy of Science*, I (1956), 53.

[24] 'Testability and Meaning' I, 440–1. When Carnap goes into more detail, the reduction sentence [R] turns out to be a special case. When the unified science seeks to introduce a new disposition predicate, Q_3, into the language, it does so through characterizing the positive and negative test cases that form the truth-conditions for application of the predicate. Carnap gives these test cases as:

[R_1] $Q_1 \rightarrow (Q_2 \rightarrow Q_3)$

and

[R_2] $Q_4 \rightarrow (Q_5 \rightarrow \neg Q_3)$,

where Q_1 and Q_4 are the antecedent experimental conditions for the test and Q_2 and Q_5 are possible results which would confirm or fail to confirm the presence of the disposition. We can see that 'Q_1 & Q_2' is a sufficient condition for application of the predicate Q_3, while '$\neg(Q_4$ & $Q_5)$' is a necessary condition for Q_3. Together, R_1 and R_2 are the 'reduction pair' for Q_3. We are told that 'By the statement of R_1 and R_2 "Q_3" is reduced in a certain sense to those four predicates', ibid. 441.

The case of [R], I said, was a special one. This is called the 'bilateral reduction sentence' and is simplified to (ibid. 442):

[R_b] $Q_1 \rightarrow (Q_3 \leftrightarrow Q_2)$

which occurs when Q_1 and Q_4, and Q_2 and $\neg Q_5$, of a reduction pair, coincide. Hence the same test can be used for both confirmation and falsification of the disposition ascription and only one reduction sentence is needed to reduce the meaning of Q_3.

which are untested. What is their truth-value? It turns out that Carnap has two responses to this: an interim response which was radical and anti-realist, and a considered response which leans back towards realism though it retains enough verificationist undertones to be objectionable to the realist. I will return to the considered response in Sect. 3.8.

Carnap's interim response to the question of untested disposition ascription is to add the clause to the reduction sentence that whenever an antecedent is not true for any time *t*, in untested cases, then it is neither true nor false to ascribe the disposition. The presence of the disposition is neither affirmed nor denied; rather, the predication of such a disposition is deemed without truth-value because it lacks any confirmed empirical content. In many ways, this is a typically verificationist solution to the problem. The disposition ascription is a proposition lacking confirmation, hence it also is an ascription lacking any significance.[25] This leaves a large area where the truth-value of the disposition ascription is undetermined, for the reduction of Q_3 is conditional upon the realization, at some time, of the antecedent test conditions of the reduction sentence. Because the meaningfulness of the term is conditional upon the truth of the antecedent test conditions, the reduction sentences are also known as 'conditional definitions'.[26]

Is it reasonable to say that it is indeterminate whether a particular vase is fragile or a particular cube of sugar soluble because

[25] Thus Carnap has to include the clause that R_1 and R_2 constitute the reduction pair for Q_3 provided it is not the case that '$((Q_1 \,\&\, Q_2) \vee (Q_4 \,\&\, Q_5))$ is not valid [i.e. true]', that is, provided at least one test case has been confirmed. Similarly, the bilateral reduction sentence [R_b] reduces the meaning of Q_3 provided that '$\forall x\, (\neg Q_1(x))$ is not valid [true]'.

[26] I have simplified the detail of Carnap's account for the purposes of exposition here. To do justice to Carnap the following should be noted. In 'Testability and Meaning' Carnap recognizes that verifiability is too strenuous a requirement for meaningful synthetic statements. He replaces this requirement with one of confirmability: 'Every synthetic sentence must be confirmable', 'Testability and Meaning' II, *Philosophy of Science*, 4 (1937), 34. Sentences are not meaningless solely in virtue of not having been, in fact, confirmed. What counts is whether such a confirmation could be achieved. Hence the statement: 'the meaning of a sentence is in a certain sense identical with the way we determine its truth or falsehood; and a sentence has meaning only if such a determination is possible' ('Testability and Meaning' I, 420). I will tackle this general principle in Sect. 3.6 but having noted the detail I will continue to oversimplify the confirmationist case for the purpose of this argument because, as we will see, the central objection to this line will revolve around completely unconfirmable cases.

they have never been tested? Could it be possible that an ascription of a disposition be neither true nor false, lacking a truth-value? Perhaps it would be too strong a thesis to attribute to Carnap that these particular examples of untested disposition ascriptions actually lack a truth-value, rather than have a truth-value that has yet to be discovered. There are, however, some cases which could be produced where indeterminacy *would* be Carnap's verdict. There is a general principle that is involved here. Our response to unconfirmed and unconfirmable cases is likely to be determined by our position on the wider issue of realism.

3.6 *Anti-Realism and the Acquisition Argument*

Although verificationism is now seldom explicitly held, some of its claims have a friend in anti-realism. The anti-realist declines to permit that all statements have truth-values. Statements for which there is no evidence or means of confirmation are among those which the anti-realist views with suspicion. Realism accepts that there is a world of objective facts, some of which we are capable of referring to with propositions. Additionally, the realist allows, while the anti-realist does not, that there may be propositions which have truth-values for which no evidence exists. The realist claims that there are truth-values for such propositions in virtue of there being facts of the matter to which propositions either correspond or fail to correspond. The anti-realist objects to such 'dissociating the truth of a statement from the availability of evidence for it'.[27]

One notable argument that has been offered in favour of the anti-realist view is the acquisition argument. The acquisition argument is a typically empiricist one, with Lockean ancestry, that explains the meaning of a term in relation to what has been acquired through experienced. Specifically, meaning is proscribed that goes beyond what can have been learnt through experience.[28] In relation to the case of dispositions, the acquisition argument provokes the following line of thought.

[27] C. Wright, *Realism, Meaning and Truth*, 2nd edn. (Oxford, 1993), 13.

[28] '[M]eaning is now being explained directly in terms of what we actually learn when we learn to use sentences in our language,' M. Dummett, *Elements of Intuitionism* (Oxford, 1977), 375–6.

What is learnt when we learn our dispositional vocabulary? The anti-realist would say that what we experience is certain events that can be observed to fall into a pattern; for example, the event of sugar being placed in a liquid is regularly observed to be followed by the event of sugar dissolving. Our experience of actual instances of dissolving sugar is then extended, as is our custom and habit, to possible dissolvings following possible placings in liquid for cases which are untested and not yet actual. To have meaning or cognitive significance a disposition ascription must be at least testable or, for reasons provided by Carnap, confirmable. Hence the following claim from Dummett: 'the criterion for the possession of a dispositional or measurable property is that of giving a certain result on subjection to a particular test'.[29] Note that the anti-realist is saying more than that this criterion is an evidential criterion. Dummett's criterion is also a criterion for whether the disposition is possessed because the essence of the anti-realist position is that evidence and ontology cannot be separated.

An unconfirmable disposition ascription is a statement for which there is no available empirical evidence which could lead to the discovery of its truth-value and thus, according to anti-realism, there is no such truth-value.[30] This can result in some counterintuitive rulings. Obviously there is wide scope for discussion of the realist/anti-realist debate but the point I make will not lead me into the details of this dispute. What I think can be said is that the ordinary use and understanding of disposition ascriptions is based on realist assumptions. There are cases which we make sense of which we would not make sense of if we were assuming anti-realism.

3.7 *Towards a Realist Theory of Dispositions*

I contend, contrary to Dummett, that true disposition ascriptions may be verification transcendent. I argue that disposition concepts

[29] *Truth and Other Enigmas*, 150. For more on Dummett on dispositions see A. Wright, 'Dispositions, Anti-realism and Empiricism', 9.

[30] I acknowledge that Dummettian anti-realism and verificationism should not be conflated completely. Dummett can allow, for instance, that Goldbach's conjecture is unverifiable without ruling it meaningless. In this case he can say that the notion of meaning as the grasping of verification-transcendent truth-conditions is undermined. Hence I restrict the claims about meaning to the conclusions about dispositions that Dummett states explicitly.

are such that their ascription makes sense and may have a determinate truth-value in cases where there is no available evidence to confirm such a truth-value. At the very least, we can say that in the ordinary way that we ascribe dispositions we are making realist claims such that the consistent anti-realist cannot participate in the discourse of disposition ascription.

An example of a statement that is arguably unverifiable and meaningless is 'everything is uniformly increasing in size'.[31] Given that spatial extension is a relative rather than absolute notion, then what makes the statement's meaning questionable is that there is apparently no difference in the way the world is whether everything is uniformly increasing in size, decreasing in size, or remaining the same. Though it is not a commitment of the present author that there are unverifiable propositions which are meaningless, clearly there is even less of a case for unverifiable disposition ascriptions being meaningless than there is for other unverifiable statements. A stronger case for an unverifiable proposition being meaningless would be where there was no possible fact of the matter, verification transcendent or not, that could determine its truth-value. The situation described by the statement and the situation described by its negation do not differ; hence it is controversial whether either has meaning. In the case of disposition ascriptions however, meaningfulness exists in unverifiable cases precisely because there is some difference, some fact of the matter, between whether the ascription is true or false, even where such a fact of the matter is inaccessible to us.

In what cases can a disposition ascription be true (or false) though such truth (or falsity) is unconfirmed or unconfirmable? Such propositions ascribing dispositions can come in different varieties with different degrees of unconfirmability. I would cite as examples from three broad categories of case:

(a) that an object could be fragile for a millennium without ever being broken,

[31] Wright, *Realism, Meaning and Truth*, 13. Taken literally, this proposition is clearly false, for given that some things observably increase or decrease their size while other things do not, any universal expansion of things cannot be uniform. I will assume for the purposes of this argument that there are no such relative alterations in the sizes of things.

(b) that an object could have a disposition that it then loses without ever having manifested it,
(c) that an object could have a disposition for which the manifestation has never yet been experienced and for which we have no concept to refer to such a disposition (and, consequently, to form truth-evaluable propositions).

I will say something more about each of these cases.

(a) Ascriptions of fragility appear dependent on the realist assumptions of those who make them and those who understand them. The breaking of a fragile object often results, directly or indirectly, in its destruction.[32] As a consequence, we can infer that fragility is usually a 'once only' manifesting disposition. The upshot of this is that the central case of fragility is one where its manifestation is *always* counterfactual. If we test the counterfactual it may well come out true—the object breaks—but then the ascription of fragility becomes redundant. However, in practice ascriptions are made, and understood, for times when the ascription is not tested. It is not the case that we are at all tempted to say that the object was fragile just for the moment at which the test was being conducted (though there are situations where this is imaginable). It is thus a small move to conclude that an object could be fragile for any length of time without actually breaking; and thus that there be a truthful ascription of fragility for the same length of time that is never put to the test. There are cases where we take precautions that a disposition is not put to the test. Just as we take precautions against explosions, we take precautions against the breakage of fragile things. Do the successes of our precautionary measures actually rob those fragile things of their fragility or merely prevent its manifestation? If such things are no longer fragile, why are we taking the precautions?

The anti-realist may challenge the realist's claim that such cases are understood. What does such understanding consist in? Is there a defensible criterion for distinguishing real from illusory cases of understanding? A full theory of understanding cannot be presented here but perhaps it need not be. It seems, rather, that we have identified a case where the very use of a disposition ascription

[32] Admittedly, some broken fragile objects go on to be repaired and may, in such a state, go on to be subjects of future ascriptions of fragility.

contains a realist assumption. To make sense of such cases is to embrace realism.

Further evidence for this emerges when a problem case is considered. Arguably there are things which satisfy the conditional in question but for the wrong reasons. Something may break when knocked, for instance, not because it is fragile but because it is in the vicinity of an explosion. An attempt to salvage the conditional analysis against cases where an object would have broken whether or not suitably knocked reduces disposition ascriptions to conjunctions of two subjunctive conditionals.[33] For example, x is fragile if and only if:

(i) if x is knocked, x will break, and
(ii) if x is not knocked, x will not break.

This conjunctive conditional analysis fails, however, and allows our realist inclinations to resurface. Some of the objects we break, though not by knocking them, may be fragile and some not; so it could be true of a fragile object that it is not knocked in the suitable way but false that it will not break. Fragile objects can break for other 'non-suitable' reasons, but the possibility of being crushed in a pneumatic press doesn't mean that a fragile object is non-fragile. Similarly, we cannot rob an object of its fragility simply by blowing it up and ensuring its breakage without it being suitably dropped. Insofar as we allow that there can be such facts about the possession of dispositions, over and above the truth and falsehood of such conditional analyses, we are realists.

(b) Cases of type (a) were cases where, the anti-realist may say, the possibility of testing exists and thus the meaningfulness of the disposition ascription is guaranteed because the associated conditional for that disposition ascription is one whose truth-value can be summoned in evidence. What, though, of cases where a disposition has been lost which was never put to the test throughout the time it was possessed? It appears that a past-tense disposition ascription can be made which never was tested and now cannot be tested.

Such cases may be quite mundane. A vase may have been fragile prior to being put through a strengthening process. A match may

[33] M. Lange, 'Dispositions and Scientific Explanation', *Pacific Philosophical Quarterly*, 75 (1994).

have been flammable prior to being doused in water. An unwound watch may have been capable of accuracy before being dismantled. In all these cases it seems that realist assumptions are revealed by our willingness to endorse the ascriptions as meaningful, having a truth-value regardless of the fact that we have no access to it.

The anti-realist could object that such ascriptions of dispositions may have, in fact, not been tested but that they were not, at the time, untestable in principle and this is the reason why they are meaningful. It is merely that no test was actually conducted when the opportunity was available. However, cases can be conceived where the possibility of a positive test result is impossible. Such cases are, for this reason, untestable and unconfirmable.

Cases where disposition ascriptions are untestable in principle have been suggested in various places by Martin. His central case is that of the 'electro-fink' where the realization of the typical test conditions for a particular disposition actually results in a loss of that disposition because of the setting up of a causal connection between the test and the disposition. We may make a disposition ascription that 'the wire is live' which could, supposedly, be analysed into some such conditional as

[A] '*if the wire is touched by a conductor then electric current flows from the wire to the conductor*.'[34]

But it could be the case that the wire is connected up to an electro-fink which is a device that detects when the wire is about to be touched and instantaneously renders the wire dead. The conditional would be false throughout the time when the wire was live. In virtue of what can we say that the ascription, being live, has cognitive significance for us? Arguably it makes sense only because we conceive of dispositional properties as real properties that are there whether or not they manifest. Thus, there is some difference in the world between a wire being live and it being dead. This is a difference which, contrary to Dummett, depends on no test or manifestation. Having or not having the disposition is itself the difference.

The electro-fink case may seem a somewhat far-fetched and artificial case but real life electro-fink cases do exist. Mellor gives the example of a nuclear power station having the disposition to

[34] 'Dispositions and Conditionals', 2.

explode but not doing so because of the various safety mechanisms that cut in just before the disposition is manifested.[35] We cannot say that because the disposition cannot manifest it is not possessed because if the disposition is not possessed then it robs the safety mechanisms of their point. Another example is Mark Johnston's chameleon sat on a green baize in the dark.[36] Assume that the chameleon is red and that being red is a dispositional property construed, plausibly, as the disposition to reflect red light in ideal conditions. Let us realize the ideal conditions by turning on the light. Is the chameleon red in these conditions? Arguably not, because realizing the test conditions alters the subject of ascription—it turns green when the light is on—and so according to the conditional analysis the chameleon was not red at any time. Put another way, according to the conditional analysis, the chameleon sat in the dark can be no colour other than green because there is no other colour it can be when the test conditions are realized. In these cases it seems that there is intuitive force in saying that the notion of the truth of a disposition ascription should be separated from the question of available evidence for such truth.

(c) The third kind of case that points to a realist conclusion is where a disposition exists which has not yet been encountered and consequently may not even be conceived.

X-ray photography is a relatively recent invention. In the time before its invention objects existed which were disposed to show up under such a test even though the test was not yet devised or even conceived. I presume that things like human bones in human bodies have always had the capacity to show up in x-rays. It would surely be quite wrong to say that such objects gained the disposition to show up under x-rays at the same time that the x-ray technique was discovered. As Martin would say, the dispositions were there, ready to go, all along. We merely needed to discover the technique that would lead to the manifestation of the disposition on demand.

In general, it seems quite a safe principle to adopt that things do not gain dispositions at the moment we gain mastery of the techniques required to stimulate their manifestations. Hence, it seems reasonable to assume that there are some dispositions of some

[35] 'In Defense of Dispositions', 167.

[36] C. Wright, *Truth and Objectivity* (Cambridge, Mass., 1992), 117.

things of which we are not aware because we have not yet discovered the way to get these dispositions to manifest. One of the aims of science itself, it could well be argued, is the discovery of previously unknown dispositions of things and ways in which these dispositions can be stimulated to manifestation.

It could be said that prior to the discovery of such dispositions there is not even a concept of what the disposition in question is. Nevertheless, something can have a disposition for which we have no concept and hence we are incapable of forming a proposition that ascribes that disposition. It would be anthropocentric indeed to say that only once we humans began to form a concept of such a disposition, and began to make verifiable ascriptions using such a concept, did the disposition come into existence.[37]

Armstrong, because of various other commitments, would object to such a claim. Martin disputes Armstrong's view that we would not allow the existence of a disposition unless somewhere at sometime it was manifested:

> There is a case that is a counterexample to Armstrong's disallowance. . . . The case is one of cosmic geographical fact concerning the spatiotemporal spread of kinds of elementary particles . . . It is supposed that there are kinds of elementary particles in some spatiotemporal region of the universe, particles different from kinds in their own region; and that the regions are so vastly distant that the many special dispositions they have for intercourse with one another never have their very special manifestations; and that nothing else in the universe, in the nature of the case, is like them that *does* have the manifestations. Yet they have causal dispositions ready to go.[38]

These examples show that the anti-realist treatment is at odds with our more liberal intuitions about disposition ascriptions. These intuitions are realist in the sense that they construe dispositions as

[37] This is not to deny that we are capable of bringing dispositions into existence in a way quite acceptable to the realist. Dennett describes a case where: 'a cardboard Jell-O box was torn in two, and the pieces were taken to two individuals who had to be very careful about identifying each other. Each ragged piece became a practically foolproof and unique "detector" of its mate . . . tearing the cardboard in two produces an edge of such informational complexity that it would be virtually impossible to reproduce by deliberate construction', D. C. Dennett, *Consciousness Explained* (London, 1992), 376. Let us call one half of the cardboard A and the other half B. Separating A and B creates two new properties and these are dispositional properties. A has the dispositional property of B-detecting; B has the dispositional property of A-detecting.

[38] 'Power for Realists', 180.

real properties that cannot be reduced in terms of conditionals. This kind of approach was considered by Carnap. Attempts were made by him and others in the tradition, however, to use this move to support, rather than be an alternative to, the conditional analysis.

3.8 *Properties, Structures, and Kinds*

The pronouncement that untested disposition ascriptions had indeterminate truth-value was Carnap's initial response to the problem. His considered response goes more in a direction that is acceptable to the realist. It still does not go far enough, as will be shown from the problems that remain, but it does move towards a view of dispositions that the realist would like to develop.

The considered response emerges when Carnap makes an attempt to further reduce the area of indeterminacy; that is, when he attempts to minimize the cases in which a disposition ascription lacks a truth-value. In doing this he reveals his view of what is responsible for the presence of dispositions. He says:

> We may diminish this region of indeterminateness of the predicate by adding one or several more laws which contain the predicate and connect it with other terms available in our language . . . In the case of the predicate 'soluble in water' we may perhaps add the law stating that two bodies of the same substance are either both soluble or both not soluble. This law would then help in the instance of the match; it would, in accordance with common usage, lead to the result 'the match *c* is not soluble,' because other pieces of wood are found to be insoluble on the basis of the first reduction sentence.[39]

On the considered view, therefore, we know that an untested match was not soluble because we know that a similar object has been tested for solubility and failed the test. Objects of the same kind have the same dispositions, Carnap stipulates, so there need be only one test on one object of the kind which then determines a disposition of the whole kind. This reduces the number of tests we need to conduct in order to make our disposition ascriptions significant or, viewed another way, it reduces the number of disposition ascriptions that are indeterminate in truth-value.

[39] 'Testability and Meaning' I, 445.

This was the first of a number of attempts to base the truth of disposition ascriptions in some property or kind membership of the subject of ascription. As such it is a move away from the conditional analysis although those who followed Carnap still could not make the move all the way.

Two similar strategies were those of Eino Kaila[40] and Thomas Storer,[41] who, independently suggested that to say that something has a disposition D is to say that it has a property P and the conditional, $F \rightarrow G$, is true of anything with that property.[42]

[40] *Den Mänskliga Kunskapen* (Stockholm, 1939), 239–40 and 'Über den Physikalischen Realitätsbegriff', *Acta Philosophica Fennica*, 4 (1941), 33–4. These works are untranslated, I have relied on the discussions in J. Berg, 'On Defining Dispositional Predicates', *Analysis*, 15 (1955), 85–9 and A. Pap, 'Disposition Concepts and Extensional Logic', *Minnesota Studies in the Philosophy of Science*, 2 (1958), nn. 4 and 7.

[41] T. Storer, 'On Defining "Soluble"', *Analysis*, 11 (1951).

[42] The details of the accounts are as follows. Where 'Fx' stands for x being subject to antecedent conditions and 'Gx' for x displaying the appropriate manifestation for a disposition D, 'Dx' is defined by Kaila as:

$$[Df_K] \quad \forall x\,(Dx \leftrightarrow \exists P\,(Px \,\&\, (\exists y\,(Py \,\&\, Fy)) \,\&\, (\forall y\,((Py \,\&\, Fy) \rightarrow Gy)))),$$

'P' being a bound property variable such that, in the case of solubility for instance, we would have the definition that x is soluble, if and only if, there exists some property P, such that x has P, and there exists some substance y with property P that is/has been/will be subjected to stimulus conditions of immersion in water (Fy), and for any such substance for which P is true and F is true, then the manifestation of dissolving (Gy) occurs.

Storer's account is similar though slightly refined. First, he introduces the time variable t into the definition, then introduces two new predicates 'WD' and '$\neg$WD'. 'WD' is true of all those things that have dissolved or will dissolve in water at some time. Thus the definition of 'WDx' is:

$$\forall x\,(WDx \leftrightarrow \exists t\,(W(x, t) \,\&\, D(x, t))),$$

that is, x is WD if and only if there is some time, t, such that x is put in water at t and x dissolves at t. '$\neg$WD' represents those things put into water at some time but did not or will not dissolve:

$$\forall x\,(\neg WDx \leftrightarrow \exists t\,\,(W(x, t) \,\&\, \neg\,\exists t\,(D(x, t)))).$$

Now Storer can introduce the property P into his full definition of solubility, which fulfils much the same role as the property P in Kaila's account. Thus solubility is defined as:

$$[Df_S] \quad \forall x\,(Sx \leftrightarrow (WDx) \vee (\exists P\,(Px \,\&\, \exists y\,(Py \,\&\, WDy) \,\&\, \neg\,\exists y\,(Py \,\&\, \neg WDy))))$$

to be read as 'x is soluble, if and only if, x is put in water and dissolves, or has some property P and there is something with property P that is at some time put in water and dissolves and nothing with this property is put in water and fails to dissolve.' Storer admits that disposition predicates involve the contrary-to-fact conditional but claims that Df_S gives a correct extensional analysis of a disposition ascription. The final clause makes Df_S superior to Df_K, for it is not open to the objection of

However, despite the formal extravagance employed in these accounts they still have the following, progressively more serious, problems.

First, the conditional employed remains truth-functional and thus incapable of expressing the fact that the connection between antecedent and consequent in the conditional is a causal one rather than merely accidental. It could still be only a coincidence that all objects, with the property P, G-ed, if F-ed; both insofar as the connection between F*x* and G*x* need not be causal but also, crucially, the connection between being P and G-ing if F-ed could also be coincidental. The realist about dispositions would be suggesting, in making a disposition ascription, that there is some property P which is actually responsible for the dispositional behaviour exhibited. The refined truth-functional account still fails to provide this. This was brought home in an example from Mackie:

> [P] might be the complex property of being cubical, transparent, and having been watched under sodium light for exactly four minutes by an albino mouse; it might be that exactly two things have [P]: this grain of sugar and that glass cube. Then if the grain of sugar is put into water and dissolves, but the glass cube is never put into water, the proposed definition makes the glass cube soluble.[43]

The second problem is that the issue of untested cases remains. What of the case where there is nothing of the same kind as *x* that has been tested for a disposition D? Arthur Pap asks us to consider such a case where:

> One might ask, for example, whether on some uninhabited planet there exists a species of metal non-existent on this planet which, like the metal species the human race has experimented with, has the disposition of 'electrical conductivity'. This question seems to be perfectly meaningful, yet it seems to be condemned to meaninglessness by Carnap's theory.[44]

Prior, Pargetter, and Jackson's 'Three Theses about Dispositions' that even if some property P is causally responsible for the manifestation of some disposition D, there remains the possibility of some particular possessing P but not D because of the presence of some 'inhibiting' property P_I. In Df_S, P*x* is a sufficient condition of S*x* only if nothing that has property P fails to dissolve in water. The drawback is that there may be no such property (hence an objection in Sect. 5.4, below).

[43] *Truth, Probability and Paradox*, 125.

[44] A. Pap, 'Reduction Sentences and Disposition Concepts', in P. A. Schillp (ed.), *The Philosophy of Rudolf Carnap* (La Salle, Ill., 1963), 561.

It seems there is a choice for Carnap between saying that the disposition ascription is vacuously true or that it is undetermined. If we say that it is vacuously true—'if F then G' is true of x because no x is F—then the paradox can resurface and x be both D and $\neg$D in virtue of also satisfying the conditional required for $\neg$D. If we say that the truth-value of a disposition ascription to an untested sample is indeterminate, then we are back with the problem in Carnap's first response that this conceptual novelty, of speaking in terms of 'same substance', was supposed to solve.

Finally, even if the first and second problems are solved by giving up the commitment to a truth-functional interpretation of the conditional, the decisive problem remains that even this Diodorean conditional need not be true for something which possesses the disposition. The reason for this is simply that the account fails to meet Martin's electro-fink counterexample. The expected conditional for a disposition could turn out to be false on every test even though the disposition ascription is true.

It is hard to see how Quine's approach to analysing dispositions, in terms of kind membership, makes any progress with these problems. Quine concedes that disposition terms are synonymous with a subset of subjunctive conditionals[45] but the subjunctive form is inadmissible in Quine's austere canonical notation for science. In response, his account is one which is closely connected to his ontological commitment that '[e]ach disposition . . . is a physical state or mechanism'.[46] For the most part we are unaware of the precise mechanisms involved in the manifestations of dispositions. Scientists will uncover these mechanisms, however, and in so doing demystify the disposition ascription: replacing the inaccurate intensional idiom with the accurate scientific explanation. As science is always developing there will always be some dispositional discourse that is yet to be eliminated.[47] In the case of the human disposition of intelligence, for instance, we are as far as we could be from a discovery of the explanatory mechanisms involved.[48] Ordinary usage does not, therefore, equate a disposition with a particular mechanism or sub-visible structure, but a precise identification is not necessary, according to Quine, as long

[45] W. V. O. Quine, *Word and Object* (Cambridge, Mass., 1960), 222.
[46] *Roots of Reference*, 10. [47] Ibid. 11.
[48] W. V. O. Quine, *The Ways of Paradox and Other Essays* (Cambridge, Mass., 1966), 72–3.

as we have faith that some structure is responsible for the dispositional behaviour, that this structure is present throughout the times when the disposition is not currently manifested, and that it makes the disposition ascriptions true in between manifestations. We are left with an understanding of disposition terms as 'promissory notes'[49] which we hope to redeem in the future as our scientific knowledge advances.

What, then, do we mean when we ascribe a disposition at the promissory note stage? To say that something is soluble is to say that it has a (for the most part unknown) structure suitable for dissolving, though it need never have dissolved. We believe an untested sample would dissolve because it has the same structure as those that have dissolved: there are structural features in common with all other soluble substances, some of which have been tested and dissolved. Much of this is acceptable to the realist. However, Quine's definition brings in something unacceptable. Using the predicate 'M' to mean 'alike in molecular structure', then 'x is soluble' would mean:

$\forall x\ \exists y$ (Mxy and y dissolves)[50]

which can be formalized and generalized to all disposition predicates:

$[\mathrm{Df}_Q]\quad \forall x\,(\mathrm{D}x \leftrightarrow \exists y\,(\mathrm{M}xy\ \&\ \mathrm{FG}y)),$

where D is the disposition predicate and FG is to be understood in Storer's sense of applying to all those things that have or will have been tested for a disposition and have or will give a positive response. Clearly the problem remains for this account that there may well exist no y which has passed the right test and is alike in molecular structure to x. There is still an assumption that some test must have occurred for the truth-value of a disposition ascription to be determinate.

3.9 *Leaving behind the Conditional Analysis*

What I have tried to do in this chapter is expose some of the empiricist, verificationist, and anti-realist assumptions that lie

[49] *The Ways of Paradox*, 72–3. [50] *Word and Object*, 224.

behind the apparent persuasiveness of the conditional analysis. The problem cases I have raised are such that they should lead us to challenge the reductive account of dispositions on offer and perhaps, ultimately, to question some of the assumptions that produced such an account. One crucial issue concerns what we intend to mean when we use disposition terms. Are we intending to ascribe properties, as the realist claims, or are we saying that certain events are possible, as Ryle and Dummett would have it? The problematic examples I have brought forward suggest the former.

The concept of a disposition is a concept of something that lies behind what occurs and what is verifiable. We accept the possibility of particulars possessing dispositions which they never manifest and we accept the possibility of kinds possessing dispositions though no kind-member has ever manifested them. The ordinary notion of a disposition permits these as significant propositions though anti-realism questions the appeal to evidence-transcendent truth-conditions.

What I suggest is the rejection of a solely conditional analysis of dispositions and that we treat them as real instantiations of properties which afford possibilities rather than just being shorthand ways of talking about certain combinations of events. The alternative view to the empiricist conditional analysis view is thus one of dispositions as instantiated properties.

However, tempting as it may be to leave this examination of the relation between disposition ascriptions and conditionals with a bold statement of realism, I think that more needs to be said. This is a question which many realists do not consider. The relationship between conditionals and disposition ascriptions is not one of equivalence but is there any relation at all; and, if not, how are dispositional properties distinguished conceptually from non-dispositional properties?

4

The Dispositional–Categorical Distinction

4.1 *The Distinction*

Most accounts have taken for granted that the dispositional is a distinct and well-defined class of terms. It is far from easy to see the justification for this. For one thing, as noted in Sect. 1.6, it is not wholly obvious what it is that dispositions are supposed to be contrasted with. Are they to be contrasted with occurrences or with states? What exactly is meant by the commonly used catch-all expression 'categorical'? What makes a property categorical as opposed to dispositional? These are some of the issues I hope to settle in this chapter.

'Categorical' means 'unconditional' and this casts doubt on the putative distinction that is being drawn because dispositions are, in a very obvious way, categorical.[1] When I say that a particular sugar-cube is soluble, I am in no way making an ascription conditionally for I am saying that it is actually soluble now, not that it could be soluble if some other condition obtained. We have seen in the previous chapter that there is an important sense of conditionality relevant to dispositions but it is the manifestation of a disposition which is conditional upon other circumstances, not the disposition itself. If this point is accepted, then disposition ascriptions are categorical—one author even refers to them as 'categorical dispositional ascriptions'[2]—and this simple point undermines many a starting assumption.

Dispositional and categorical properties have been discussed as if the difference between them was a question not even worthy of

[1] See Sect. 3.2. My thanks to Ullin Place for stressing the importance of this point to me (in correspondence).

[2] J. Bricke, 'Hume's Theory of Dispositional Properties', *American Philosophical Quarterly*, 10 (1975).

consideration. Commonly this difference could only be suggested by bringing forth paradigm cases from each class rather than by offering a general criterion for distinguishing them. What it is that makes a particular property type belong to one or the other category is left unsaid, a situation I aim to remedy.

A statement of the dispositional–categorical distinction is important because the question of a conceptual distinction between the dispositional and the categorical is a problem that is prior to the question of any ontological distinction. In other words, it makes no sense to try to establish a property-dualistic ontology for the dispositional and the categorical, as some have attempted, if there is not even a conceptual distinction between the two. If there is no conceptual distinction, then there is no ontological one either. There is, however, a job for ontology if the conceptual distinction can be maintained, for whether there is a parallel ontological division remains an open question in such a case.[3]

The importance of this question is justifiable at a theoretical level but it will be useful to show also the significance of the problem in some practical cases. One simple piece of evidence is that not all property terms divide neatly into either the categorical or dispositional classes. Consider the property term 'molten' which we apply to rocks and metals under extreme heat. Is this a dispositional or a categorical term? Solely using pre-theoretic intuitions, a simple answer is not easily found. A good case can be made for classifying it as either. In ascribing the property of 'moltenness' we are saying what the rock or metal is actually like now but we also say a lot about how it could behave in various situations: that it will burn, flow, glow, and so on. Colour concepts are notoriously difficult to place in respect of the dispositional–categorical divide. Are colours to be understood as categorical properties of objects, or are they 'merely' dispositions to cause sensations in perceivers?[4]

[3] Addressed in chs. 5 and 7, below.

[4] A problem noted by E. J. Lowe, 'Sortal Terms and Natural Laws', *American Philosophical Quarterly*, 17 (1980), 254 and G. Dicker, 'Primary and Secondary Qualities: A Proposed Modification to the Lockean Account', *Southern Journal of Philosophy*, 15 (1977). For some arguments concerning colour's dispositional or categorical status see F. Jackson and R. Pargetter, 'An Objectivist's Guide to Subjectivism about Colour', *Review of International Philosophy*, 41 (1987), where it is argued that colour is non-dispositional, and J. Harvey, 'Challenging the Obvious: The Logic of Colour Concepts', *Philosophia*, 21 (1992), where it is argued that it is dispositional.

These are just a few examples where we have difficulty in categorizing certain property ascriptions and, correctly, this need not be considered a proof that the dispositional–categorical distinction is in doubt. In other cases there seems no doubt about the distinction. Without hesitation, we classify 'fragile', 'soluble', 'elastic', and 'magnetic' as disposition concepts and 'triangular', 'broken', and 'of molecular structure *xyz*', for any specified molecular structure, as categorical concepts. These are some of the paradigm cases of each class that are often cited but a problem is that until we can provide a precise statement of the distinction we don't know whether these classifications are a result of prejudice and familiarity or principled distinction.

The accepted view is that ascriptions of dispositions have a special relation to subjunctive conditionals that non-disposition ascriptions do not have.[5] In Chapter 3, it was shown that this relation between disposition ascriptions and true conditional statements is not one of equivalence, which would permit the reduction of dispositions to complexes of events named in the antecedents and consequents of those conditionals. An alternative and perhaps more common view is that conditionals are entailed by disposition ascriptions. If we accept '*x* is fragile', we must, on this view, also accept some stronger-than-material conditional such as 'if *x* is dropped or knocked in a certain manner, then *x* will break'. If we accept '*x* is soluble in water', then we must also accept 'if *x* is immersed in water, then *x* dissolves'. Dispositions such as elasticity are slightly different. If we accept '*x* is elastic', then we must accept a set of conditionals: 'if pulled, then stretches', 'if released, then contracts', 'if pressured, then bends', 'if dropped, then bounces', and so on. Entailment of a set of conditionals makes elasticity, or any other disposition, a multiply manifested or 'multi-track' disposition. This form of the orthodox view, that conditional entailment is the criterion of the dispositional, is one to which I will offer support. This support must be qualified, however. I will argue that the connection between dispositions and conditionals is indirect and mediated by the functional roles that we ascribe when we ascribe dispositions. This means that the conditional entailment by a disposition ascription is itself conditional upon certain implicit

[5] e.g., Goodman, *Fact, Fiction and Forecast*, 34ff.; Ryle, *Concept of Mind*, 123; and Quine, *Word and Object*, 222.

or explicit assumptions made for the ascription. These assumptions can never be complete enough to guarantee the non-trivial truth of any particular conditional though a conditional is a way in which functional roles can be articulated. Such conditional entailment that there is will also be shown, therefore, to be dependent upon the understanding of dispositions as real properties that support the possibilities that a conditional describes.

Two challenges show that the orthodox view needs revision. The two challenges to the orthodoxy are: first, that disposition ascriptions do not uniquely entail conditionals; second, that not even disposition ascriptions entail conditionals. These two challenges come from opposite directions and are clearly inconsistent: one says that conditional entailment is in all property ascriptions; the other says that it is in none. A defensible statement of the distinction will have to show that both positions are wrong, however, and that there is a set of property ascriptions that do entail some such conditionals in the appropriate way and a set of property ascriptions that do not.

4.2 *First Objection: Disposition Ascriptions Do Not uniquely Entail Conditionals*

The defender of the dispositional–categorical distinction, it has been suggested, must hold that categorical property terms do not have the same relation to subjunctive conditionals that dispositional terms have. The orthodox view has been challenged by Mellor. The challenge is simple but powerful, though not, I shall show, conclusive. Mellor argues that, in addition to disposition ascriptions, even categorical property ascriptions entail conditionals. He says:

> Take the paradigm, molecular structure—a geometrical (for example, triangular) array of inertial masses. To be triangular is at least to be such that if the corners were (correctly) counted the result would be three. Inertial mass entails only subjunctive conditionals specifying acceleration under diverse forces.[6]

[6] 'In Defense of Dispositions', 171. In this argument Mellor is following Popper, 'The Propensity Interpretation of the Calculus of Probability', and Goodman, *Fact, Fiction and Forecast*.

If Mellor is correct, the dispositional–categorical distinction should not be made with reference to conditionals for entailment of conditionals is not a unique feature of the dispositional. Analogous to:

[A] 'x is soluble' entails 'if x is in water, then x dissolves'

is:

[B] 'x is triangular' entails 'if the corners of x are counted, then the result will be three'.

We can assume that the holders of the distinction will accept triangularity as an unproblematic categorical property, for if triangularity is not an unproblematic categorical property then it is difficult to see what else is. If even the supposedly unproblematic 'very paradigm' of a categorical property ascription entails a subjunctive conditional, then surely all categorical properties can be assumed to do likewise. If conditional entailment is the sole basis of the distinction, as many take it to be, the distinction is thus collapsed.

4.3 *An Unsuccessful Reply*

But is Mellor's argument a good one? Elizabeth Prior has argued not and has attempted a defence of the distinction against Mellor's attack.[7] Prior argues that there is a distinction between the conditionals [A] and [B], above, that Mellor is missing.

According to Mellor, triangularity can be given the following analysis:[8]

[1] *'x is triangular' entails 'If the corners of x were (correctly) counted the result would be three'*,

which differs from

[2] *'x is triangular' entails 'If the corners of x were counted the result would be three'*,

[7] 'The Dispositional/Categorical Distinction', repeated in her *Dispositions*.
[8] The following paraphrases from Prior.

which Prior claims is 'a closer analogue than [1] of the (clearly) dispositional [3]':

[3] *'x is fragile' entails 'If x were (suitably) dropped, x would break'.*

But [2], we are told, is false, for the corners of a triangle could be miscounted—though granted, this would require extreme ineptitude—or, in a possible world, a systematic deception could occur such that the result regularly was four or five (the result may on occasion be three where a systematic deception coincides with a miscount).

The difference between the false [2] and the true [1] is that [1] contains the qualification that the counting be done 'correctly'. What is intended by this clause? Mellor states, unequivocally, that the 'correctly' refers to 'how the counting is done, not to whether it gives the result three'.[9] If it did refer to the result rather than method of counting, it would be the trivially analytic conditional:

[i] *'If the corners are counted such that the result is three, then the result will be three.'*[10]

So that there is no confusion, Mellor even tells us how to do the counting in a non-trivially correct way: we put the corners in a one-to-one correspondence with an initial segment of the sequence of positive integers, count them only once each, and the highest number is the result.[11]

Prior disputes Mellor's specification of the term 'correctly'. The method of counting could be correct, she insists, and the result still not be three: in our possible world where a systematic deception occurs. Thus, claims Prior, the 'correctly' in Mellor's analysis must refer to the result of counting rather than the method of counting. If 'correctly' refers to the result, then the conditional is, as Mellor agrees, trivially analytic. But Prior claims that genuine dispositional conditionals, such as [3], are not trivially analytic. Therefore, the dispositional and categorical can be distinguished on the basis of the latter, but not the former, entailing only trivial conditionals. Thereby the distinction is preserved.

[9] 'In Defense of Dispositions', 171, n. 35.
[10] D. H. Mellor, 'Counting Corners Correctly', *Analysis*, 42 (1982), 96.
[11] Ibid. 97.

We can see that the distinction is not preserved, however. Prior has claimed that

[1] '*x is triangular' entails 'If the corners of x were (correctly) counted the result would be three*'

and

[3] '*x is fragile' entails 'If x were (suitably) dropped, x would break*'

differ. This involves at least two contentious claims, both of which it is wise to reject: first, that the '(suitably) dropped' in [3] is significantly different from the '(correctly) counted' in [1]; and, second, that a correct method must always guarantee the correct result.

What is the warrant for treating [1] and [3] differently in these respects? Prior's justification seems to be given when she says:

Here we do have a conception of being suitably dropped that can be explicated without reference to the notion of fragility. I am not saying that 'being suitably dropped' is easy to explicate but I am saying that it is plausible that it can be explicated without reference back to fragility. This is precisely what cannot be done with 'being correctly counted' in the case of triangularity.[12]

The danger, which Prior is trying to down-play, is that the 'suitably dropped' could be taken as referring to a result of dropping rather than a method of dropping; that is, 'suitably dropped' meaning 'dropped such that breakage occurs', yielding the trivial:

[ii] '*If x is dropped such that breakage occurs, then breakage occurs*',

which makes everything fragile (the danger perhaps goes further than the evident triviality, for it seems a non-trivial, but reasonably plausible related point, that just about anything could break if dropped in a non-trivially suitable way; that is, dropped from a great height on to a hard surface).[13] The problem with this is obvious: if the 'suitably dropped' produces a trivially true conditional, like that produced by 'correctly counted', then the dispositional–categorical distinction, once again, collapses.

In what sense is Prior optimistic that a non-trivial explication is possible for fragility? The explication must presumably be one that

[12] 'The Dispositional/Categorical Distinction', 95.

[13] Though sponges, tennis balls, and h's are exceptions.

is specific as to the method of dropping, for example, one which is specific as to the force of the impact administered, which one assumes will be a product of the height of the drop and the hardness of the surface upon which the object lands. There is admittedly a big problem of vagueness in the word 'fragile' here, but we must agree that an object is fragile when the force of impact necessary for breakage is 'comparatively low' (call the conditional with this clause 3*). Even if we did work with such a notion of fragility the problem would not be solved. Mellor's explication of triangularity could only ensure the correct result by referring to the result of counting, not to the method; doesn't the same apply to 3*? We cannot ensure that every impact of a force above the minimum will result in breakage unless we build such a clause into 3* rendering it also trivially analytic. 3* does not ensure breakage because the *point of impact* may affect the way the stresses are absorbed through the fine structure of x, such that one point of impact of above minimum force will result in breakage while a different point of impact of the same force will fail to result in breakage. We could try to build in a clause so that 3** can handle such a possibility but even then there is a further problem: two identical impacts, at identical points, to structurally identical objects, need not produce the same result if some indeterministic microstructural event has a causal influence, and we have every reason to believe that such chance events do occur in our world.

As in the case of Mellor's explication of triangularity, therefore, we cannot guarantee the truth of the consequent of our conditional unless we refer to the result rather than method of testing. This is no big deal, however. There seems no real justification for Prior's claim that dispositions always necessitate their manifestations; in our probabilistic world, why not say that they just highly probabilify them? Similarly, it is highly probable that we will get a result of three when we count the corners of a triangle, but we cannot guarantee this; accidents will happen. The systematically deceptive world, where corners are regularly counted incorrectly, could just as well be a world where breakages regularly fail to occur to fragile objects, in which case we must follow Mellor and try to find some other method of testing for triangularity or for fragility.[14]

Prior's argument is rendered hopeless because the conditionals

[14] 'Counting Corners Correctly', 97.

entailed by ascriptions of triangularity and fragility are not distinguishable. Her point that the subjunctive in [1]

'*If the corners of x were (correctly) counted the result would be three*'

sometimes fails to be true, is matched by the failure of the subjunctive in [3]

'*If x were (suitably) dropped, x would break*'

to be always true. The only way in which both subjunctives could be guaranteed to be true is if they both make reference to the result of their respective tests rather than the methods of testing employed.

The attempted distinction fails. Prior has another argument for the distinction but unfortunately, as well as being far from clear, this rests upon what was controversial about the first argument: that the 'correctly' clause of 'counting' refers to the result rather than method.[15]

Given that we accept that all dispositions need not be 'surefire'—their stimuli need not necessitate their manifestations—we can pronounce Prior's defence of the distinction a failure.

4.4 *Loose Living: A Problem for Mellor*

Prior's defence has failed so we have ruled out one putative specification of the distinction but this doesn't mean that the distinction cannot be salvaged. Mellor's argument does not rule out all defences and the debate has been left unsatisfactorily resolved. We may still be able to find a different criterion of demarcation.

To reinforce this view, it is useful to note an ambiguity in Mellor's position. Mellor concentrates on the role of conditionals in the meanings of disposition ascriptions but the relation of these conditionals to dispositions is not entirely clear. Are we to interpret Mellor as claiming that disposition ascriptions entail true conditionals or that disposition ascriptions are equivalent to true conditionals or something else?

Mellor claims that he is not attempting to give a full analysis of disposition concepts: his 'defence of dispositions' is that what was thought objectionable about them—their conditional entailment—

[15] 'The Dispositional/Categorical Distinction', 95.

is a feature shared with what we accept to be respectable properties. He says: 'My strategy will be to show the offending features of dispositions to be either mythical or common to other properties of things; just as loose living is no prerogative of the unmarried and so is no proper basis for discriminating against them.'[16]

We can concede to Mellor, therefore, that both dispositional and categorical property ascriptions do, in some way, entail conditionals without drawing the conclusion that the dispositional and categorical are indistinguishable. Indeed, Mellor seemed reluctant to draw such a conclusion himself and it is no part of his argument in that paper. It seems, therefore, that Mellor has given no compelling reason at that place why we should not concede most of his argument but defend the distinction on some other grounds. I should note before I attempt to provide such grounds, however, that in a later paper on the subject Mellor says something quite different about the status of categorical properties. He says: 'Elizabeth Prior tries to sustain the myth that there are non-dispositional properties of things, i.e. properties that support no subjunctive conditionals.'[17] This is an evident abandonment of the dispositional–categorical distinction but it seems to have no more justification than the adoption of the definitional *fiat* associated with the conditional analysis:

if the ascription of x entails a (subjunctive) conditional, then x is a dispositional property

and this is not in accord with the aims of the earlier paper. In his first paper, Mellor warns against discrimination, in an analogous case, because not only single mothers are loose living. The claim in the second paper, that all property ascriptions are dispositional because all are conditional-entailing, seems rather like the inference in the analogous case that we are all single mothers because we are all loose living.

4.5 *Preliminaries to the Functionalist Theory*

Having followed the Mellor–Prior debate to its unsatisfactory conclusion, it is apparent that a statement of the dispositional–categorical distinction is no simple matter. Little has been provided,

[16] 'In Defense of Dispositions', 157.

[17] 'Counting Corners Correctly', 96.

so far, by way of positive thesis. I now attempt a better defence of the distinction against Mellor's attack along lines that avoid all the problems we have encountered thus far.

In defending the distinction I make assumptions that are shared by C. B. Martin, who is nevertheless soon to become an opponent (Sect. 4.7). These assumptions are justifiable on the conceptual and intuitive grounds that were developed in the previous chapter: dispositions are actual, intrinsic states or properties rather than 'bare potentialities' and that to say something is now soluble is to say something about what it is like actually rather than something about possible future events. The view that dispositions are not actual properties is, as has already been suggested, either a confusion of the disposition with its manifestation or a reduction of dispositions to events as in the conditional analysis. The claim that dispositions are intrinsic properties is the claim that they are instantiated properties which inhere completely in the object of ascription and, thus, can be possessed in the absence of their stimulus conditions. Hence, it is a mistake to think that something can be elastic only relative to the occurrence of an actual stretching. This feature was first noted in the case of fragility in Sect. 1.2.

It is possible, of course, that a disposition be possessed in between its manifestations or possessed though never manifested at all and consequently it is possible that it be possessed though we may never know it. A length of tape can become magnetized at a time t_1 giving it a disposition to play back the film *Casablanca* in suitable circumstances (i.e. if in a video recorder connected to a television). This tape may be destroyed at time t_2 without ever manifesting its disposition but it seems eminently plausible that it did actually have the disposition between t_1 and t_2. Thus it is accurate to speak of dispositions as *actual* properties which 'support' potentialities: the potentialities of their manifestation. It is counter-intuitive, to give another example, to say that the human skeleton gained the disposition to make x-ray impressions at the moment that x-ray technology became sufficiently advanced to create images and that it did not have the disposition before that moment. Present-tense disposition ascriptions, therefore, say something categorical about the present.[18]

[18] Disposition ascriptions need not, of course, be stated in the present tense e.g. it could have been, between times t_1 and t_2, that x had disposition D. There are also be future-tense disposition ascriptions. These cases are not essential to the argument, which can easily be extended to them.

Now consider the ascription of the so-called categorical properties. It is quite difficult to find, anywhere in the literature, a specification of what exactly is intended by 'categorical property' and it is symptomatic of the vagueness of the distinction accepted thus far that a number of different things are usually put forward as examples of the categorical. The list of ontological categories includes occurrences, episodes, shapes, structures, and molecular substructures. Most obviously, but least committally, a categorical property is any property which is not a dispositional property—a contrast which cries out for the addition of some positive content. Categorical property ascriptions, let us take it, say something about what is actual but it is demonstrable that they also say something about what is possible, and for this we need only refer to Mellor's analysis of triangularity. To say something is triangular is certainly to say that it is triangular, categorically, now; but it also has implications for what is likely to happen, if the corners are counted.[19] What we need to show is that there is something about the conditional entailment by dispositional properties which is different from the conditional entailment by 'occurrent', 'structural', or 'categorical' properties. The difference is not, as Prior thought, to be found in the entailed conditional itself; rather it is in *the type of conditional entailment*. This is something that is revealed most easily when we consider exactly what it is we are ascribing when we ascribe a disposition. I now introduce an account of disposition ascription that I am going to be developing in more detail in the remainder of the book.

A disposition ascription, I take it, is a functional characterization of a property: it is the classification of a property according to its functional role. It is to characterize a property according to what effect it will produce in a particular circumstance or, in other words, the role that the property plays midway between a stimulus/manifestation pair of events. Thus we say '*x* is water-soluble' when *x* has some property which, in certain circumstances, causes dissolution when in water.[20] There may be other ways of referring to

[19] What is at issue is the possibility of something happening and this need not be stated in the future tense e.g. it could have been possible, between times t_1 and t_2, that if P, then Q. As in the previous note, this point is not important to the main issue.

[20] What is meant by 'in certain circumstances' will be considered in Sect. 4.9. This account of the semantics of disposition ascription has a basis in R. Harré, 'Powers', *British Journal for the Philosophy of Science*, 21 (1970), 85 and E. Fales, *Causation and Universals* (London, 1990), 195.

this property but we refer to this causally efficacious property as 'water-solubility' when we wish to characterize it functionally.

We say nothing about the categorical description of this property when we make a disposition ascription. What is not involved in, or necessitated by, a disposition ascription is a specification of the non-dispositional characterization, be this in terms of shapes, structures, or states. A disposition is the disposition it is purely in virtue of the satisfaction of functional criteria, which are distinct from the criteria for its so-called categorical nature. Thus, to give a disposition ascription is to say something about what a thing can do but to say nothing about how it does what it does. To give an example that may illuminate this abstract statement: what is common to all thermometers, as we saw (Sect. 1.2), is a functional essence—an ability to react consistently on exposure to varying temperatures and to supply a display thereof—which can be realized by any number of structurally differing mechanisms. 'Thermometer' is, for this reason, best construed as a disposition term rather than a categorical one.

All that is necessarily entailed by the ascription of a categorical, as opposed to a dispositional, property, I take it, is the existence of some actual shape, state, structure, or property. So far, this does not exclude dispositions, for as we saw, they also are actual. However, with the categorical, nothing is entailed, purely conceptually, about what causal or functional role such a shape, state, structure, or property will play in its interactions with other things. Hence it is not possible to derive anything about the causal role of a property, analytically, from the meaning of a categorical property ascription.

A categorical property ascription may be necessitated, according to a number of positions,[21] by every disposition ascription, but what specifically the categorical 'base' property is, for any disposition, is an a posteriori matter. The categorical base of a disposition is, according to these accounts, the categorical property, or complex of categorical properties, that underlie a disposition. The basic line I am promoting here is that it is an empirical matter, for any disposition, what its categorical base is. Inversely, it is

[21] As in the position of Armstrong's realism, *Materialist Theory of the Mind*, 85–9 or my own identity theory, S. D. Mumford, 'Dispositions, Bases, Overdetermination and Identities', *Ratio*, 8 (1995).

similarly an empirical matter, for any categorical base, what disposition(s) it supports.

Our knowledge of the causal roles of the various categorical properties is a complex matter of scientific investigation. As such investigation has been conducted on a large scale already, there is the temptation to take the causal role as analytically connected to the categorical property when actually our knowledge of the causal role of a categorical property is imbedded in a sophisticated physical theory. Such a causal role is not, therefore, part of the meaning of a categorical ascription, as it is with disposition ascriptions. Dispositions are causal-role occupiers by conceptual necessity; categorical properties are causal-role occupiers, but only a posteriori causal-role occupiers in the sense that what particular causal role a categorical property occupies is something that has to be discovered.[22]

We can now make explicit the conceptual distinction between the dispositional and the categorical, notwithstanding the second objection to the distinction, to be considered in Sect 4.7:

The Distinction

Disposition ascriptions *are ascriptions of properties that occupy a particular functional role as a matter of conceptual necessity and have particular shape or structure characterizations only a posteriori.*

Categorical ascriptions *are ascriptions of shapes and structures which have particular functional roles only a posteriori.*

Note should be made of a possible difficulty with the distinction thus stated. It may be thought that certain apparently categorical terms do contain their causal roles as part of their meaning. While it seems convincing that one can understand what it is to be

[22] We can, at this stage, provide a solution to an outstanding problem in Sect. 1.2, namely what is the difference between the 'abstract' dispositions discussed by Aristotle and the more typical 'concrete' dispositions. The distinction is now obvious. In the case of concrete dispositions the functional role by which we characterize the disposition is a *causal* role; with abstract dispositions the disposition has a functional essence but one which is not causal. Thus Aristotle made the distinction quite well when he said that the primary kind of potency, akin to our concrete dispositions, were 'originative sources of change', unlike the 'powers' of abstract entities which can originate no change. What exactly the functional role of a non-causal entity amounts to is, of course, an intriguing question, but one I will not attempt to answer at present.

triangular without understanding the causal role triangularity plays for the objects that possess it, it seems less convincing that some theoretical terms such as 'lepton' and 'meson' contain causal roles as no part of their meanings even though they have the appearance of categoricality. In reply to this it needs to be understood, first, that leptons and mesons are theoretical objects, rather than properties, and, second, that their causal roles are part of their meaning because they are objects whose essential properties are dispositional. In Sect. 1.2, I gave the example of a thermostat as an object with a dispositional essence. It seems that we should be dispositional essentialists also about certain fundamental theoretical entities, such as electrons, where it is an object's dispositions to behave that make it the thing that it is.[23] The response, therefore, is to be steadfast in the face of this objection. Any apparently categorical term which does have its causal role as part of its meaning is to be understood as a dispositional term in virtue of that fact.

4.6 *The Functionalist Response to the First Objection*

Regarding Mellor's problem of the status of triangularity we can say the following: it is a conceptual truth that if x is a triangle, then x has three corners, but whether or not we can count them correctly is not a conceptual truth and it cannot therefore be derived by necessity from an ascription of triangularity. Though it is true, almost always, that when we count the corners of a triangle we get the result three, this truth depends upon the contingencies of the laws of nature—Prior's systematically deceptive world is, after all, a possible world—we therefore know the truth of the subjunctive (in [1], above) only a posteriori.

In the case of [3] ('x is fragile' entails 'If x were (suitably) dropped, x would break') there is a conceptually necessary connection between the property ascription and the conditional. Given Martin's electro-fink objection, which I tackle in the next section, the connection between a disposition ascription and a conditional is obviously not as straightforward a connection as

[23] See B. Ellis and C. Lierse, 'Dispositional Essentialism', *Australasian Journal of Philosophy*, 72 (1994).

some empiricists and anti-realists have supposed. It will be seen that such conditionals are true only when certain circumstances obtain. However, the functionalist asserts that there is a conceptual connection between being fragile and having the causal role of breaking upon being suitably dropped when such circumstances obtain. In virtue of this, there is a conceptual connection between being fragile and the conditional 'if drops, breaks' because the conditional form is a way of expressing a functional role. Admittedly, the truth of this conditional will itself be conditional upon 'background conditions' but I will leave the question of such background conditions until later.[24]

It seems, if the analysis I have given is correct, that the conceptual distinction between the dispositional and the categorical is defensible against this first attack if we distinguish between the modal notions behind the bare concept of 'entailment'. In the one case the entailment is conceptual and hence logical, in the other case the entailment derives from an a posteriori discovery. Stronger-than-material conditionals are 'entailed' by both dispositional and categorical ascriptions but in the case of dispositions the relation is a priori as opposed to a posteriori in the case of categorical ascriptions. Structure or related concepts also are 'entailed' by both disposition and categorical ascriptions, but the entailment is conceptual in the case of the categorical, not so in the case of the dispositional.

Something more needs to be said about so-called natural necessity, its origin and nature. I accept plausibility in one Humean claim: that it is logically contingent which type of events follow events of another type. I call this 'the contingency thesis' and will refer to it again in Sect. 10.8. The thought it gives expression to is that the way the world works could have been otherwise to the extent that the processes in nature are not logically dictated or bounded but are 'only' what is the case. A common reaction to this thought is that what dictates the processes which actually occur, in

[24] A rival way of describing functional roles is with Ramsey sentences, as detailed by D. Lewis, 'Psychophysical and Theoretical Identifications', *Australasian Journal of Philosophy*, 50 (1972). The utility of explicating functional roles in this way is not being challenged by my present concentration on the relations between disposition ascription, conditionals, and background conditions. I concentrate on conditionals because the Ramsey sentence statement is not the best way of illuminating the problem of interfering background conditions, to be discussed in what follows.

this world, are laws of nature which have application, universally, to various kinds. The laws of nature could have been otherwise. This can be expressed in possible world discourse as the claim that there are possible worlds where the laws of nature are different from the laws of nature in the actual world. An impact such logically contingent laws will have on the dispositions of objects is that it is a contingent matter what dispositions issue forth from what categorical properties. The Humean would have it, for instance, that given the categorical properties of things, there is no logical necessity in what dispositions those things possess. What dictates a thing's dispositions is the contingency of whatever the laws of nature happen to be. Hence there is some sense in the possibility of other worlds in which objects with those categorical properties have different dispositions from those they have in the actual world, though the Humean may be reluctant to state the point in these terms because of their aversion to reifying possibilities. I will not be explaining the contingency thesis in terms of the existence of contingent laws of nature, though I will be supporting the contingency thesis, and in due course I will offer an alternative explanation. The upshot for the dispositional–categorical distinction is that the causal role of a property cannot by deduced logically from a categorical ascription, by conceptual analysis alone, because what the causal role of a categorical property is will be discovered only upon the examination of the contingent facts.

As I see it, I am following ordinary usage in this account of the dispositional–categorical distinction, though I am making the distinction more explicit. It is not controversial, I maintain, that dispositions are ordinarily taken to have a functional essence and I will be outlining and defending this understanding of dispositions further in Chapter 9. I am quite happy, returning to our original disputed cases, to follow ordinary usage with a term like 'molten' where it was noted that the distinction seemed to blur. I have proposed, though, a criterion of distinctness which would seem to set up a sharp boundary. My explanation is that 'molten' is a single property term which is used to refer to both a dispositional and a categorical property: a possibility which is not ruled out by my account. Similarly, colour terms can be treated in one case as referring to the categorical correlate, or primary qualities, of the disposition to cause subjective experiences; in another case referring to the disposition itself. Given such examples, it seems that we are being too hasty if we treat our doubts over which class to place such terms in as conclusive evidence that the classes themselves are indistinct.

However, having introduced a functionalist theory that answers Mellor's attack, it needs to be shown now that this account of what it is to be a dispositional property can be defended against the second attack on the distinction.

4.7 *Second Objection: Not Even Disposition Ascriptions Entail Conditionals*

C. B. Martin has made a contribution that has a significant impact on the issue of the dispositional–categorical distinction.[25] His conclusion that disposition ascriptions are not equivalent to stronger-than-material conditionals is largely to be accepted and, indeed, is one I have endorsed and used against the conditional analysis. However, the argument also raises a problem. Should Martin's argument be accepted, then there would be no basis for the dispositional–categorical distinction. I cannot see how this could be a result that Martin would welcome because if the distinction is destroyed then the prospect of replacing a conditional analysis of disposition ascriptions with the 'first-order dispositions or powers' he recommends[26] seems empty of content. The problem concerns what makes any property *P* a dispositional rather than a categorical property and the threat posed to the distinction is quite different from the first objection that I rejected.

The orthodox view, that I have offered some indirect support to, would have it that *P* is a disposition ascription if and only if *P* entails a subjunctive conditional (by conceptual necessity) but Martin's argument suggests that this is neither a necessary nor a sufficient condition for a disposition ascription. So what it is that makes the ascription of *P* the ascription of a power remains mysterious. This question is one that Martin needs some answer to, for he needs some backing to the claim that there are dispositions. If there is no such account of the distinction, then not only is the widely held assumption that dispositional properties are

[25] 'Dispositions and Conditionals.' [26] Ibid. 7.

problematic correct, categorical properties are equally problematic for the same reason; that is, there will be no account of what it is for a property ascription to be a member of such a class.

I now present a more detailed account of Martin's argument. A conditional analysis of disposition ascriptions, as we saw, has it that the ascription of any disposition predicate D is equivalent to a subjunctive conditional such as 'if F was the case, G would be the case'. This is a completely reductive account of dispositions: disposition ascriptions are merely allusions to possible sequences of events, they are not ascriptions of properties to things. In agreement with Martin I take it that dispositions *are* real properties that can be gained or lost even though there could occur no events which betray their presence and it is for this reason that we must be at pains to emphasize the distinction between a disposition and its manifestation.

Martin considers the disposition ascription

[a] *The wire is live*

and the proposed conditional which is meant to be its *analysans*

[b] *If the wire is touched by a conductor, then electrical current flows from the wire to the conductor.*[27]

But it turns out that the proposed *analysans* [b] is neither necessary nor sufficient for the disposition ascription [a]. This is demonstrated by considering the case where the wire is connected to a machine called an electro-fink which detects when the wire is about to be touched by a conductor and can instantaneously render a dead wire live or a live wire dead.

In the first case (i) imagine that the wire is not touched by a conductor and the wire is dead. Because of the electro-fink, the conditional [b] is true; that is, if the wire was to be touched by a conductor, then current would flow from the wire to the conductor. Hence we can see that [b] is not logically sufficient for [a]; for in this case [b] is true but the wire is dead, that is, [a] is false.

In the second case (ii), where the electro-fink is operating on a 'reverse cycle', [a] is true—the wire is live—but [b] is false. In this case the electro-fink makes it such that if the live wire was to be touched by a conductor it would instantaneously become dead;

[27] 'Dispositions and Conditionals.' 2.

hence current would not flow from the wire to the conductor, as is required for the truth of [b]. This shows that the conditional [b] is *not logically necessary* for the truth of [a], which confirms the result that [b] is not the *analysans* of [a]. Generalized, the result is that because (i) the conditional could be true and the disposition ascription false and (ii) the disposition ascription could be true and the conditional false, disposition ascriptions are not equivalent to subjunctive conditional propositions. This is a conclusion anyone attracted by realism about dispositions ought to welcome.[28]

Note that Johnston's chameleon example (Sect. 3.7) works in the same way. Does the fact that the chameleon would appear green to an observer whenever they turned on a light to look tempt us to say that it is green when there is no light on? Arguably not. If we accept the analysis 'x is green if and only if (if x is observed in ordinary light, x appears green)', we should, for consistency, answer in the affirmative.[29]

Rightly, one way out of this problem has been blocked. We cannot say that a disposition ascription is reducible to a conditional with an 'other things being equal', or *ceteris paribus*, clause added to ensure its truth. Martin revives Goodman's argument that such a clause would have to rule out all possible relevant factors to the truth of [a].[30] This would mean that the new conditional [b′] would have to include the intelligibility of [a] to be itself intelligible, for all relevant factors would be all factors relevant to the truth of [a]. Hence [b′] would not reduce [a].

However, the conclusion that is welcomed by realists also raises a major difficulty for them. The orthodox view has it that what makes any ascription a disposition ascription is not that it is equivalent to a conditional proposition but that it entails one. Martin's account would mean that not even disposition ascriptions

[28] David Lewis has a response to Martin, 'Finkish Dispositions', *Philosophical Quarterly*, 47 (1997) which he says has considerable agreement with mine. Certainly Lewis has a conclusion that is welcome for the realist though he tries to incorporate this into a reformed conditional analysis. The basic claim is that x has a disposition D iff it retains the causal basis for that disposition. This causal basis could well be identical with the disposition, as Lewis acknowledges, but he also says that he wants to be non-committal on this question.

[29] For further scrutiny of the 'conditional fallacy' see R. K. Shope, 'The Conditional Fallacy in Contemporary Philosophy', *Journal of Philosophy*, 75 (1978).

[30] Goodman, *Fact, Fiction and Forecast*, ch. 1.

entail conditionals, for the disposition ascription could be true and the conditional false, as in Martin's case (ii). This means that an account of the dispositional–categorical distinction is lacking from Martin's position. I presume he must think that there is such a distinction, for he is happy to admit dispositions as a meaningful class. Rather, it is the articulation of disposition ascriptions in conditionals that is his target. He says that such conditionals are 'clumsy and inexact linguistic gestures to dispositions and they should be kept in that place'.[31] But what, without such conditionals, makes it such that it is a disposition that is gestured at and not a categorical property? How can the use of conditionals be avoided here? Even Martin admits that 'Statements ascribing causal dispositions or powers are *somehow* linked to (strict or strong) conditional statements.'[32] Perhaps Martin wishes to say that dispositions are just the causal powers we try to gesture at with conditionals, so we may as well gesture at them directly instead of via conditionals. It seems that every property must be a causal power in this sense, however, for every property must be causally potent (as a necessary condition of its very existence?).[33] This makes it appear highly unlikely that to avoid a statement of the dispositional–categorical distinction we can simply assert that some properties are causal powers and others are not.

4.8 *An Unsuccessful Reply*

My aim is to support an account of dispositions which preserves the dispositional–categorical distinction while accepting dispositions as real properties, as Martin wishes. This can be achieved with the functionalist account.

Dispositions have functional essences in the sense that what it is that makes any particular disposition the disposition it is is the functional role it plays and what it is that makes any property a dispositional property is that it has a functional essence. For example, what it is that makes all things soluble is that they are apt for dissolving when in a liquid. This is a conceptual point: it

[31] 'Dispositions and Conditionals', 8. [32] Ibid. 2, Martin's emphasis.

[33] This is a contentious principle but one I find appealing. I will say something in its defence in Sect. 6.2.

couldn't be that the property of fragility could have occupied the causal role of solubility instead (if 'the property of fragility' is understood as a non-rigid designator picking out, at every world $_w$, whatever property at $_w$ has the same causal role as fragility at the actual world). Thus, what it is in virtue of which any particular thing is soluble is that it has a property, or property-complex, that will be causally efficacious of dissolving in a liquid when the conditions are right. This is a conceptual truth: any property-complex that causes dissolving when in liquid can be given the functional characterization 'soluble'. Which particular 'categorical' properties also occupy these causal roles, if any, is a contingent matter. Similarly, what it is that makes anything a thermostat—the essence of a thermostat—is the causal role that the object will play, namely triggering a switch when a certain threshold temperature has been reached. Which combinations of categorical properties combine to have this functional role is a contingent matter; hence any conditional that follows from such property ascriptions does not follow as a matter of conceptual necessity but only contingently insofar as their causal roles are dependent upon the contingent laws of nature—the facts of which are already assumed in the case of a disposition ascription. Therefore dispositions, *qua* functional properties, are to be contrasted with any properties that do not occupy their causal roles as a matter of conceptual necessity and this is sufficient for the contrast of dispositions with all the usual things they are typically opposed to: occurrences (because they are not properties), structures and shapes (because they do not occupy particular causal roles by conceptual necessity), and so on.

What is essential to a disposition—its identity condition—is its functional role. This raises the question of how a property's functional role relates to a conditional because Martin's example shows it to be an erroneous move to go straight from functional role to conditionals.

It may be thought natural and inoffensive to articulate or describe a property's functional role in the form of a conditional. However, to do so without precautions is to fall into Martin's trap. If we accept Martin's account, then no conditional follows by conceptual necessity from a disposition ascription. The items playing a crucial, frustrating, role here are the 'background conditions'. For any true disposition ascription, the possibility of background conditions that would prevent manifestation is left open.

In an earlier paper I attempted to deal with the problem of interfering background conditions in a way that I now think is unsatisfactory.[34] This strategy begins by drawing the sustainable distinction between two types of background condition:

α-conditions: being *conditions that prevent the manifestation of a disposition though the disposition itself remains*, for example: lack of oxygen prevents a struck match from lighting though it remains flammable; the lack of a mate prevents a man from breeding though he remains fertile; placing a vase in a sturdy glass case prevents it from being broken though it remains fragile.

and

β-conditions: being *conditions that prevent something from having a disposition*, for example: a match being wet stops it being flammable; a zero or low sperm count stops a male from being fertile; a strengthening process stops a vase from being fragile.[35]

When a disposition ascription is made there is no accompanying explicit statement of what α- and β-conditions are to be ruled out but in making a disposition ascription we are making an implicit claim that none of the β-conditions for that disposition, whatever they are, is currently operative.

A disposition cannot be possessed when there are β-conditions but it may be possessed when there are α-conditions, and this is where Martin's case (ii) gains its credibility. The problem for any conditional-entailment account of the dispositional is that α-conditions render conditionals false though not the associated disposition ascriptions. I tried to defend a view that although no straightforward disposition is entailed by a disposition ascription there is nevertheless a qualified conditional that is entailed, namely, causing G, if F, provided there are no α-conditions. I then said that although this looks like the attachment of a *ceteris paribus* clause to the conditional, Martin's objection to *ceteris*

[34] 'Conditionals, Functional Essences and Martin on Dispositions', *Philosophical Quarterly*, 46 (1996).

[35] These two types of background condition explicitly articulate a distinction made by R. S. Woolhouse, 'Counterfactuals, Dispositions and Capacities', *Mind*, 82 (1973), 559.

paribus clauses applied only to accounts that attempted to be reductive of disposition ascriptions to conditionals and this account was not attempting to be reductive.

However, I now think that we have reason to be dissatisfied with the above response to the problem because of the difficulty in filling out the *ceteris paribus* clause, especially given the definition of α-conditions above. If α-conditions are conditions that prevent the disposition from manifesting itself, then when expanded the conditional with the α-condition attached becomes 'if F, G, unless not-G', which renders the conditional vacuously true of everything. Attaching an α-condition exclusion is like saying that when the antecedent is realized then the consequent will be realized unless something stops it. That cannot do.

4.9 *The Context of Ideal Conditions*

The functionalist best copes with the electro-fink problem by sticking firmly with a realist theory of causation and not attempting to analyse a functional role in terms of conditionals.

Clearly some connection between dispositions and conditionals cannot be denied. Even Martin says that they are connected 'somehow' but he is puzzled by what the 'somehow' connection is. The connection I suggest is as follows. A disposition ascription is a statement that something occupies a particular functional role. This raises the question of which functional role we are attributing when we make a particular disposition ascription. A functional role is specified in terms of what causal consequences a property or state will produce in response to what antecedents. A natural way of stating these is in conditional form. However, we have seen that the truth of such a conditional cannot be guaranteed because of the possibility of some interfering factor preventing the disposition manifestation. This makes apparent that it is not specifically a problem with conditionals that is at issue. The problem is a more general one of how we can state the function played by a particular or property. How can we say that something has a function to do φ if there is always the possibility of some interfering factor that will prevent it from doing φ? It seems that we have exactly the same problem here so we are wrong to think that the problem is purely one of conditionals.

Note that this problem infects realist accounts such as Martin's as much as it does an empiricist account. The realist says that disposition ascriptions are ascriptions of real powers. This leaves unanswered the question, 'power to do what?' The problem of background conditions means that the realist cannot say what it is that a power is a power to do. The electro-fink itself is subject to the problem. The electro-fink's function is to make a live wire dead or a dead wire live whenever a test situation is initiated. It can only do this when conditions are right, though. There is the possibility of some interfering background condition that stops the electro-fink from doing what Martin says it does. How, then, do we say what the electro-fink can do in certain circumstances when it may not always be the case that it does that thing in those circumstances?

The possible interfering background conditions cannot be excluded in a finite list that is appended to the conditional. This is because there is no finite list that could name all such possible conditions in which the manifestation is prevented. To state that the excluded background conditions are any conditions which interfere with the disposition manifestation is to render the conditional trivial.

The minimal claim of a disposition ascription is that a particular *can* do something. Given that this condition is met by most things, however, something more will usually be meant. What is usually implied by a true disposition ascription is that there are background conditions, let us call these 'ideal conditions', in which such manifestations do follow from the stimulus. A disposition ascription thus invokes the following '*conditional* conditional':

[Df_i] if C_i, then (if Fx, then Gx),

where C_i represents the ideal conditions, F and G represent stimulus and manifestation respectively, and both conditionals have subjunctive force.

The immediate danger of saying this is that it too, like the α-condition stipulation, can be true only if it is trivial. There are ideal conditions in which any particular can manifest any reaction if 'ideal conditions' is interpreted trivially. Certainly this line would gain support from Prior's remarks on the 'incompleteness' of dispositional predication. Every substance is soluble at a high enough temperature or if in the right solvent.[36] Similarly every physical object is fragile at a low enough temperature so it seems, arguably, there are some conditions in which if F, then G, for any disposition, could be true of any property, and if these conditions are those appealed to as the ideal conditions in Df_i, then the account is trivial.

However, I would argue that Df_i can be defended under a non-trivial interpretation for C_i. Under this interpretation, what count as ideal conditions are determined by the context of the disposition ascription. To say something is soluble is to say it will dissolve, in liquid, in a context relative to the ascription. The ascription in the actual world is relative to actual world conditions. It is also relative to actual world conditions that can vaguely be understood as 'normal'. Disposition ascriptions are made for a reason. The ideal conditions that facilitate the manifestation of the disposition are thus expected to be ones which are not realized only in exceptional circumstances. If such ideal conditions were exceptional, relative to the context of the disposition ascription, then there would be little utility in making the disposition ascription. Whoever made such an ascription would be, if not strictly speaking a deceitful ascriber of dispositions, at least an uncooperative or misleading one. In making an appropriate and useful disposition ascription I am saying that, in ordinary conditions for the present context, if a particular antecedent is realized, a particular manifestation usually follows. This licenses someone to *expect* the manifestation when the antecedent is realized even though it is possible that the expectation be disappointed. If it is disappointed, then some explanation of why it was may be available. If no explanation is available, it may still be thought that there was some reason, some interfering background condition, why the manifestation did not occur. It is even possible that the manifestation fails to occur, for a different reason, every time the disposition is put to the test. Though this would be a surprising case, revision of the disposition ascription would not automatically follow even in such extraordinary circumstances.

This is not to deny the possibility of making a disposition ascription that will hold only in unusual conditions. This is the reason why I do not speak merely of 'ordinary conditions' instead of ideal ones. In many cases the context of the disposition

[36] *Dispositions*, ch. 1.

ascription may suggest ordinary conditions but not in every case. A scientist may state that under such-and-such extreme conditions, a sample may be expected to exhibit such-and-such behaviour. They may, for instance, be theorizing about how certain objects will behave at extremely low temperatures approaching absolute zero, or about how objects might be expected to behave when entering a black hole, or about the behaviour of elements which can be instantiated only in an artificial environment. The point is that in such cases the exceptional conditions will be fixed by the context of the ascription. That these are unusual background conditions will have to be flagged for the disposition ascriptions to have a point. Of course, even in these cases, the scientist concedes that something may interfere and prevent the expected manifestation.

How can this point, about the context relativity of relevant ideal conditions, offer an explanation of Martin's electro-fink objection? Where the wire was live but the associated conditional appeared to be false, the position I have advanced suggests the following reply. The conditional [b] (If the wire is touched by a conductor, then electrical current flows from the wire to the conductor) is part of the understanding of what it is to be live. However, although it remains true that there is a conceptual connection between the disposition ascription and the conditional, the conditional is true only against the background of ideal conditions. Such ideal conditions regularly are realized and manifestation of the disposition regularly does occur. Some property is there, being live, and this property has a functional essence of enabling the flow of electrical current through a conductor. The possibility of this ability being possessed but also being prevented from manifesting is accepted but the fact that we feel a need for some explanation of this failure is evidence that we detect a disparity: we feel that there must be some interfering condition at work. Ordinarily the manifestation would be expected but there is an explanation of why it does not occur. Our further expectation is that if this interfering background condition is removed, the disposition will manifest itself once again.

One more point needs to be clarified before moving on. Is a disposition ascription, like a categorical one, embedded in a contingent physical theory and thus impossible to distinguish from a categorical ascription? The only apparent difference between the two is that the contingent conditions concern ideal conditions for manifestation, in the case of a disposition ascription, rather than laws of nature. Thus, while the causal role of a categorical property is known only once contingent laws of nature are known, the causal roles of dispositional properties are known only once contingent ideal conditions are known, so the cases are alike.

This is not quite an accurate account of the situation concerning what is known a priori. What is known a priori with a disposition term, because it is part of the meaning, is the causal role that is being ascribed when an ascription of that term is true. What is to be found out, through empirical investigation of contingent situations, is the specification of what conditions are normal and abnormal, given the context of that ascription, and what would constitute the ideal conditions for the manifestation of the disposition. With categorical terms the situation is reversed insofar as the causal role of the ascribed property is what is to be found empirically because the causal role of a categorical property is dependent upon contingent laws. A way in which this difference is demonstrated clearly is as follows. Were the laws of nature to change such that things took on different causal roles from the ones they have at present then the reference of categorical terms could remain unaltered; that is, terms like 'triangular' could pick out the same properties. Given that the causal roles of categorical properties had altered, however, it would be different things which were soluble, fragile, and so on. 'Soluble' would still refer to the same causal role of dissolving in liquid though there may have to be new investigation into the best conditions for its manifestation. The causal role of something that was triangular, however, would not be known until new empirical studies into the matter were complete.

4.10 *The Defence of Realism*

The essential point that needs to be kept in mind in the defence of a realist theory of dispositions is that conditionals are only, as Martin has said, inexact gestures towards the properties that sometimes make them true. Over the last two chapters, disposition ascriptions and conditionals have been shown to be more and more loosely connected if we understand dispositions to be real properties involved in real causal interactions. In Chapter 3, I argued that dispositions could not be reduced to the class of events. In this chapter I have tried to develop a view that although dispositions can be said to entail some form of conditional, these conditionals can only be specified loosely because they are only a by-product of the functional characterizations of properties that we are making when we are speaking of dispositions. In this functional characterization, what is vague and loosely defined is not the stimulus and manifestation conditions. These are quite tightly defined and follow from the disposition ascription by conceptual necessity. But what is left unspecified, and dependent upon the context of the ascription, are the ideal conditions in which such a conditional is true. These cannot be stated in full. For any set *S* stating a list of such conditions, there is always the possibility of some interfering condition *C* which renders the entailed conditional false though *S* holds in all stated respects. The set of ideal conditions can thus only be assumed to be unremarkable for the context of the disposition ascription.

What I hope to have achieved in this chapter is a plausible account of the dispositional–categorical distinction that makes some reference to conditionals but is nevertheless consistent with realism. This is the sort of account that I think is lacking in Martin's negative programme. I would hope that this account could be seen as the positive programme that actually supports that kind of realism about dispositions that Martin has recommended.

5

Property Dualism

5.1 *Irreducibility as an Ontological Thesis*

In the last two chapters I was concerned with the conceptual distinction between the dispositional and categorical. I attempted to reinforce the notion of disposition ascription being a distinct class of property ascription distinguishable from other sorts of property ascription and event description. Dispositions are best understood as properties functionally characterized and they can be contrasted with either events or any property that is not functionally characterized.

With the present chapter the discussion turns away from the conceptual problems. I shift the focus onto the ontological problems that the legitimizing of disposition ascription brings. By looking at the thesis I have called property dualism, I aim to raise problems that will occupy me until Chapter 10.

The most pressing of these problems, that raises itself instantly, concerns whether the world is really composed of two distinct types of properties: the dispositional and the categorical. Unlike many previous combatants in the debate, I distinguish this as a specific and significant ontological claim because I want to bring into sharper focus the issue of whether we should say that there are separate dispositional and categorical properties at work when a disposition manifestation is produced. In distinguishing such a position I need to go beyond the classificatory division of ontological positions that was introduced by Mackie: the division of theories into phenomenalism, realism, and rationalism.[1] Mackie's taxonomy of theories is, I contend, blind to an important issue that divides adherents to those positions. Mackie is not alone in being insensitive to this issue and at least one author whom I accuse of being a property dualist[2] has objected that I am forcing

[1] *Truth, Probability and Paradox*, 121 ff. [2] Ullin Place, in correspondence.

him into a pigeon-hole and that in some cases he takes a monist stance though a dualist one in others. The division I make between dualist and monist theories is, however, nothing more than the analogue of the dualism–monism division in respect of the mind–body relationship. If any writer claimed that on this issue they were a dualist in some cases and a monist in others, then we would reply that if they are a dualist in any instance, they are a dualist overall, for they allow the possibility of at least one substance (property, event) from an ontologically distinct category. The same must be said in the case of the dispositional and non-dispositional: if the existence of two distinct and irreducible kinds of property is allowed, then a theory that allows it is a dualist theory.

How does a thesis of property dualism arise? Given the establishment of a conceptual distinction between the dispositional and the categorical, then the prospect of a similar division in ontology arises quite naturally, though it could not have done so otherwise. What has been shown in the last chapter is that disposition ascriptions are conceptually irreducible and only because of this can the possibility be entertained that such irreducibility exists in ontology as well. This is certainly a conceivable possibility, notwithstanding the comments I will be making below in Sect. 5.8. If, on the other hand, it had been found that no conceptual distinction was sustainable, then the possibility of an ontological division would not have been viable either; for a conceptual reduction of the dispositional would entail an ontological reduction. Given that there has been no such conceptual reduction then the dualism–monism issue remains open; to be settled by engaging the ontological issues.

My concern in this chapter will be to examine the arguments that have been put forward for property dualism as an ontological thesis. Some of these arguments I will attempt to dismiss straightaway but others will be shown to have an initial degree of credibility and so will set an agenda that any anti-dualist alternative will have to address. These will have to be put off for further consideration because answers can only come when the plausibility of the anti-dualist alternative is understood.

However, after stating the arguments, and conceding the force of some of them, I will show why I think that a serious dualism of properties is a position we are motivated to argue against. I will argue that there are some counter-intuitive consequences to property dualism and that a plausible alternative, that does not have these counter-intuitive consequences, is desirable. First, though, I begin by attempting to articulate the dualist claim and so clarify exactly what is at stake.

5.2 *The Dualist Claim*

The dualist claim is that there are distinct dispositional and categorical properties inhabiting the world in the same way that dualists in respect of other types of property may argue that there are distinct mental and physical properties or distinct temporal and non-temporal properties.[3] Given that the division between dispositional and categorical ascriptions has been sustained, is it not a natural move to say that disposition ascriptions are true in virtue of the presence of dispositional properties and categorical ascriptions are true in virtue of the presence of distinct categorical properties? The dualist is thus claiming that there is a fundamental bifurcation of reality and that to have a dispositional property is a different sort of thing, in some substantial way, from having a categorical property. In respect of objects and substances, the dualist position urges that in addition to a thing's categorical properties, such as shape, texture, and microstructure, it also has a set of dispositional properties that are a different type of property to categorical properties and that in some way account for the thing's behaviour. There is a claim of some real division in reality here: it is not merely that there are two ways of talking about a thing's properties; there are, rather, two different types of property that we can talk about.

This is a position which I am going to argue against. I think that it views the nature of properties and property ascriptions in completely the wrong way and I will be offering an alternative based on examination of the ontological issues. However, I want to show the initial appeal of the dualist position and so first I consider some of the claims that I think contain evidence of dualism. The following can be offered as examples:

[3] An example of a temporal property would be being 10 years old.

[W]e cannot wholly reduce causal characteristics to non-causal characteristics by correlating the former with the persistence of a certain type of internal structure;[4]

[P]otentiality is irreducible to actuality; potentiality is not identical with one or a series of present or possible observables;[5]

One can make neither type-type nor token-token identifications of dispositions and bases. However it is true that whenever an item possesses a disposition D it also possesses a physical property P which is the basis of the disposition;[6]

[D]ispositions have a 'categorical irreducibility', as it is impossible to explain them away in terms of other categories such as space, time, form, process, material, property, etc. For suppose that the exact shape and size of an object were known, the shapes and sizes of all the constituents, along with a list of these facts at every time. We would still know nothing about how or why the object would change with time or on interactions. Still less could we predict how it would respond to a new experimental test.[7]

An initial point of clarification is that I understand all of these statements to be claims about dispositions, as explicated thus far, even though they may not use that term explicitly. Hence I am taking potentialities and causal characteristics, along with other categories such as powers, abilities, tendencies, and propensities, to be dispositions. Subtle distinctions may be possible but the account of dispositions I develop is general enough to cover all these items.

Three more substantial aspects of the dualist position need to be brought into focus:

1. *Causal and categorical bases*

The notion of a causal, microstructural, or categorical base needs to be explained. What is usually meant by such a thing is something like the following. For any disposition possessed by an object or substance, there is, as a matter of empirical fact, or it is inductively probable that there is, some property that we might think of as the base property or basis of that disposition. The basis *b*, of

[4] Broad, *The Mind and its Place in Nature*, 435.

[5] Weissman, *Dispositional Properties*, 14.

[6] Prior, *Dispositions*, 81.

[7] Thompson, 'Real Dispositions in the Physical World', 69.

any disposition *d*, is generally understood to be that property, or property-complex, in virtue of which the object or substance has *d*. Armstrong expresses this point in terms of *b* being the 'truth-maker' of an ascription of *d*. It is frequently made in terms of *b* being the explanation of *d*.

There are, however, divergent views about the way in which *b* could be constituted. Mellor says that the basis of one disposition d_1 may be another disposition d_2 and the basis of d_2 may be yet another disposition.[8] If this view is correct, then a disposition and its base could be distinct though there is only one type of property: all properties could be dispositional. Typically, the dualist makes the claim that the basis of a disposition is non-dispositional, however: it is substantially a different sort of property altogether. Place refers to the basis as microstructural. An example of such a basis would be the molecular structure of sugar which explains why it is soluble. Some bases of dispositions may well be macrostructural properties, though, as in the case of the base property for the disposition of a ball to roll in a straight line when hit, which can be explained in terms of its shape and centre of gravity. As I mentioned in the last chapter, properties such as these are among those which in general are understood as categorical, and indeed are still understood as such on the revised definition of 'categorical' that I have developed. Thus there is the general way of referring to such bases as 'categorical bases' of dispositions. This is the way in which Prior, who gives one of the clearest statements of property dualism, refers to such bases.[9]

2. *The denial of identity*

One thing I take to be a necessary condition for a dualist position is the denial of identity between the disposition *d* and its base property *b*. Such a denial of identity is not sufficient for dualism, however, because Mellor denies the identity of a disposition and its base but still has a property-monistic ontology: all bases, as he understands them, are dispositional.

It will be seen that one of the major erroneous assumptions in the debate, to date, is that the question of identity is the only question we must answer to decide between dualism and monism.

[8] 'In Defense of Dispositions', 174. [9] *Dispositions*, ch. 5.

Something else is required for dualism: the acceptance of different categories of property. Mellor accepts only one category of property—the dispositional—with various dispositional properties standing in various explanatory relations to each other. However, a denial of identity between *different categories* of property is a more serious ontological matter that leads to the dualistic position I am describing. If the disposition and the base are of different classes of property and if identity between such classes is denied, then we have two ontologically distinct kinds of property. What a typical monist position is at pains to deny is that having a disposition and having an appropriate categorical base for that disposition is the having of two properties of fundamentally different kinds.

3. *The relationship between dispositions and their bases*

If dispositions are not identical with their bases, then what is the relationship they have to them? There must be some relation because the dualist has to provide some explanation of why it is property b_1, rather than b_2, that is the base of disposition d_1. Identity of b_1 and d_1 would be one explanation but the dualist has ruled out the relation of identity.

One likely kind of relation that the dualist is to appeal to is a causal one. There are, however, two different ways in which the causal relation could hold:

[A] First, it could be thought that the categorical base of a property causes a particular to have a dispositional property which in turn causes that particular, in ideal conditions, to *G*, if *F*-ed. Something like this causal structure is argued, for instance, by Weissman[10] and Place (see Sect. 5.5, below). The dualist in this case is arguing that once the categorical property has caused a particular to have a disposition, that disposition is then sufficient to cause manifestation events unaided by the categorical base.

[B] The second possibility is that the causal base of a disposition is itself the property which is sufficient for the production of the manifestation event in suitable circumstances. Thus, the disposition does not cause the disposition which in turn causes the manifesta-

[10] *Dispositional Properties*, 187f. Weissman divides properties into static and dynamic, real dispositions or potentialities being the latter, and then says that the static determine the dynamic.

tion: the categorical base causes the manifestation, in ideal conditions, directly. This is the causal structure that is advocated by Prior.

A causal relation is not the only possible one, however. There may be a supervenience claim where dispositions supervene on their categorical bases. This would be explicated in traditional style as there being no change in a disposition of a thing without there being a change in its categorical properties and two particulars which are alike in all their categorical properties being alike in all their dispositional properties.

Hence the typical dualist position will be one where dispositions have categorical bases which are distinct from, and of a different type of property from, their disposition but are related to their disposition(s) and in virtue of this relation some explanatory relation exists also.

Having identified some of the key concepts and issues, I move on to arguments for the distinctness of a disposition and its base.

5.3 *Explanatory Asymmetry*

Let us start by examining a short and simple argument towards a dualist position. O'Shaughnessy has advanced the following argument in favour of the distinctness of the dispositional and the categorical.[11] It is an argument against the alleged identity of dispositions with their categorical bases but it obviously points to the broader conclusion that dispositions are identical to no categorical properties at all, hence dispositional and categorical properties are distinct. The force of the argument trades on an alleged asymmetrical explanatory relation: categorical bases explain dispositions therefore they cannot be identical with such bases, for something cannot be identical with what is its explanation.

I will break down the assumptions at work and show why the conclusion is not to be supported.

First, there is the plausible claim that the categorical base is the explanation of the disposition. Given what I have said above, on categorical bases, it might even be thought that this claim is true by definition given that the categorical base property is identified on the

[11] B. O'Shaughnessy, 'The Powerlessness of Dispositions', *Analysis*, 31 (1970), 5.

basis of being the property that has this explanatory relation to the disposition. What, though, does such a categorical base explain? I noted that there were two candidates for the causal relation between a base and its disposition which presumably any explanatory relation is dependent upon. The base could cause [A] the disposition or [B] the manifestation of the disposition. The question to be asked of O'Shaughnessy's claim now becomes whether the base is alleged to explain the disposition or the manifestation of the disposition?

If we say [A] that the categorical base explains the disposition, then why conclude from this alone that the disposition must be distinct from this base? We may well argue that this is just as good an argument for identity as it is for distinctness. Certainly identity between the disposition and the categorical base is consistent with the explanatory relation that O'Shaughnessy suggests, for a possible way in which the categorical base could explain the dispositional property is that it is identical with the dispositional property. There is thus no reason to make the assumption that the explanatory relation rules out the possibility of identity. Further, it seems, if the categorical and dispositional were not identical, then something else would be needed for there to be an explanation, namely, something that accounts for the necessary connection between the categorical and dispositional properties.

If we said, instead, that [B] the categorical base explained the manifestation of the disposition, then the problem for the argument is that the possession of a dispositional property can also be offered as the explanation of its manifestation. Prior to the discovery of the categorical base of the disposition, the manifested behaviour usually is explained by the ascription of a disposition, even though Quine may refer to such an explanation as 'pre-scientific'. The point is, however, that if a dispositional property could fill the same explanatory role as a categorical property then, again, there is a position that is consistent with the thesis of identity because properties that filled the same explanatory role could well be identical.

The Quinean view is that dispositional explanations should be construed as promissory notes for the full explanations that will come once we have taken into account all the empirical evidence.[12] Here the disposition term is construed as a place-holder that designates indirectly whatever the categorical property explanation

[12] *Roots of Reference*, 14.

is. If so, then we come very close to saying that the disposition just is this categorical property described in a 'pre-scientific' way, which again, is a conclusion of identity.[13]

However, even if we put these objections aside and concede that dispositions are explained by non-dispositional categorical bases, I do not see how it decides anything conclusive because there is not the kind of explanatory asymmetry that O'Shaughnessy would need. As Franklin notes, a categorical property always entails a further dispositional property.[14] Thus, if we want an explanation of why a particular substance has a certain molecular structure, we could explain it in terms of the dispositions of its constituents to form bondings of a certain kind and to remain bonded in that way. An explanation solely in terms of some smaller-scale structure will be an inadequate explanation of the larger structure, for there will be no explanation of why a substructure of a particular type is disposed to form a certain type of larger macrostructure.

The situation is, therefore, that though categorical properties provide explanations for dispositions, categorical properties are also explained by dispositions; hence neither category uniquely explains the other.[15] O'Shaughnessy's argument can thus be countered on the grounds that both types of property ascription can be used in an explanation of the presence of a property of the other kind so there is not the explanatory asymmetry that the argument claims.

5.4 *Variable Realization and the Swamping Argument*

I come now to a more serious problem. The following argument for the distinctness of a disposition and its 'causal base'[16] appears in Prior, Pargetter, and Jackson:

[13] This is not the only possible interpretation of Quine's claim. See Sect. 8.1.

[14] 'Are Dispositions Reducible to Categorical Properties?'

[15] One possible problem with this picture is that if dispositions explain structures and structures explain dispositions, then there is a regress of explanation. This problem will be addressed in ch. 9.

[16] A clarificatory note on this position: Prior, Pargetter, and Jackson just speak of the distinctness of a disposition from its 'causal base' without committing themselves to the nature of this base. However, if the categorical is synonymous with the non-dispositional, then these causal bases must be categorical (as Prior states explicitly in her *Dispositions*) for, as we shall see, dispositions are declared to be causally impotent with respect to their manifestations.

> [T]here is the consideration that it is empirically plausible that certain dispositions have different causal bases in different objects. Suppose in particular that the causal basis of being fragile in some objects is molecular bonding α, in others it is crystalline structure β. (It does not matter whether this is plausible for being fragile, because it can hardly be denied that this happens with some dispositions. And in the case of the disposition of being fit in Evolutionary theory there are a multitude of different bases.)
>
> We cannot say both that being fragile = having molecular bonding α, and that being fragile = having crystalline structure β; because by transitivity we would be led to the manifestly false conclusion that having molecular bonding α = having crystalline structure β.[17]

Another example is produced by Mackie of the water-absorbency of cloth having two bases: water is absorbed between the individual fibres and in the fibres themselves.[18] Again exploiting the analogy with philosophy of mind, we can call this claim the claim that dispositions are variably realized by a number of categorical bases; therefore they must be distinct from those bases.

Prior, Pargetter, and Jackson say that the variable realization of dispositions is a plausible empirical hypothesis and that this is all that they need for their conclusion of distinctness; they don't have to show variable realization occurs in any particular case. It may be enough to show that there is the possibility of variable realization (see Sect. 5.6, below), but let us consider for a moment whether the hypothesis is actually true.

Mackie's example of the different bases of absorbency is one that looks suspicious, for the two bases of absorbency may actually be reducible to one. The holding of water in the fibres may just be a smaller-scale version of the holding of water between the fibres. Fibres are made of smaller fibres. If water is absorbed because there are small gaps, of varying sizes, within the structure of the cloth, then this, and only this, is the causal basis of absorbency. We therefore have some reason to be suspicious of this particular variable realization claim for water-absorbency.

In one place, I considered the possibility that we might have a principled reason for thinking that all such variable realization cases could be explained away as errors of the same kind as Mackie's.[19]

[17] 'Three Theses about Dispositions', 253.

[18] *Truth, Probability and Paradox*, 152.

[19] 'Dispositions, Supervenience and Reduction', *Philosophical Quarterly*, 44 (1994).

His example may be spurious but a lot more work would be needed to show that all are spurious and I now think that such an attempt would be mistaken.[20] Many different things are fragile, including objects made of different materials. Could it be shown that there is but a single categorical property that bases all the different instances of fragility? Sugar and salt, despite differing in molecular structure, are both soluble so surely solubility is variably realized. Consider the disposition of being breakable: so many different things are breakable that it seems highly implausible that one, and only one, categorical property could be common to all such things. I am willing, therefore, to endorse the claim that different categorical structural properties are capable of realizing the same disposition.

The argument for dualism is not completed though, even if the plausibility of variable realization is granted. In the analogous case in the philosophy of mind, a simple switch to a thesis of token–token identity results in the accommodation of both monism and variable realization; indeed variable realization is typically used to motivate token–token identity theories. Prior thinks the case of dispositions differs and that the same move cannot be made.[21] This is an issue I will consider and contest in Chapter 7.

However, something further can be said. Even if variable realization is accepted and token–token identities for the case of properties is not allowed, something more still needs to be said for the type of property dualism I am considering. Variable realization is consistent with the position that a disposition and its causal base, though not coextensive, are nevertheless different properties of the same type. What would need to be added was support for the claim that dispositions are a different type of property from the properties that constitute causal bases. Prior's dualism does have a reason for this kind of view, however. She thinks that dispositions are distinguished by being causally impotent whereas the properties that constitute causal bases are potent. This leaves her with a curious position that I will discuss in due course. The opposite view may be thought, on conceptual grounds, to be more attractive: dispositions may be understood as exclusively the causally potent properties.

[20] Nor is it necessary, however, for the ontology that I advocate.
[21] *Dispositions*, ch. 6.

Prior, Pargetter, and Jackson give another argument for the distinctness of a disposition and its causal base. They go on to say that:

> [T]here is the difficulty that even if there is only one causal basis of fragility, say, bonding α, it may happen that although all fragile objects have α, some objects that have α are not fragile. This would be the case if there were an internal structural property S which swamped the effect of having α. Any object which has α but not S is fragile, every fragile object has α, but those objects which have both α and S are not fragile. This seems a perfectly possible state of affairs, but not if fragility = having α.[22]

I call this the *swamping argument*.

Certainly it seems empirically plausible that there could be a property F which caused a disposition manifestation in one instance but if it occurred with another property F_s, its effect would be prevented and thus the disposition not possessed. No actual examples are given by Prior *et al.* but let us concede its possibility.

A concern about this argument, however, is that it works only by begging the question against disposition–categorical base identities. If the swamping argument is to be given its strongest statement, it must not be assumed, for any example, that the property in question is non-identical with the disposition but it appears that this is assumed. We are asked to imagine a property that all fragile objects possess. Presumably this cannot be *any* property possessed by such objects, such as being greater than 0.01 grams in weight, because this might be a property that is only accidentally present in all such cases. The property monist will say that the property in question is supposed to be a plausible candidate for the categorical property that is identical with fragility; and the most plausible candidate is the categorical property that is causally relevant to breakage upon being dropped. This is why categorical bases are sometimes called, and so called by Prior, Pargetter, and Jackson, 'causal bases' of dispositions. This categorical property would, therefore, be one that excluded the swamping property F_s; that is, $\neg F_s$ would be a necessary condition of any property, or property-complex, that was responsible for breakage upon being dropped. α, in the example cited by Prior *et al.*, is a property that

[22] 'Three Theses about Dispositions', 253.

does not exclude F_s, therefore it is not a possible candidate for the property identical to fragility and, in assuming a property non-identical with fragility to be the candidate property identical with fragility, the argument begs the question against an identity theory.

I think, therefore, that it looks bad for the swamping argument unless its proponents can restate it in a way that avoids this charge. The best that can be made out of it is perhaps the claim that *as a matter of fact* there is no single property F that can fill the place of α and which does exclude all swamping properties. In such a case, however, the argument amounts to nothing more than the variable realization claim of the previous argument and so adds nothing more to the case for dualism.

5.5 *Place's Dualism*

Place argues what I take to be a dualist line and offers three reasons why dispositions are not identical with their microstructural bases. I will take these reasons one at a time and explain why we should accept none of them.

(a) *Differences of category*

First, Place argues:

> Two descriptions cannot be descriptions of one and the same thing if there is a difference of category between the kind of thing picked out by one description and that picked out by the other. In the case of the alleged identity between dispositional properties and their basis in the microstructure, both descriptions are descriptions of properties, but they are descriptions of properties of different kinds. Dispositional properties are modal properties, they consist in their possible future and past counterfactual manifestations. The microstructural properties of an entity on the other hand are categorical, which, of course, is why Armstrong who finds modal properties offensive wants to reduce the dispositional to the microstructural.[23]

This argument cannot be accepted because the description of dispositions as modal properties, consisting in their possible future

[23] D. M. Armstrong, C. B. Martin, and U. T. Place, *Dispositions: A Debate* (London, 1996), 60.

and past counterfactual manifestations, is not an acceptable one. Place's view of dispositions here is close to that of Ryle's, namely a conditional reductionist view of the kind I argued against in Chapter 3. Place himself admits that disposition ascriptions are categorical, in a sense, in that they ascribe actual properties. Because of this, however, the difference in category, in the way Place understands it, cannot be accepted. Dispositions are more than just 'modal properties'. A disposition consists in more than just its possible manifestations; it consists in the possession of a property or properties classified according to the functional role played by such a property or properties. The full details of this alternative to the conditional analysis have yet to be presented but once they have been it will be clear that the difference of category objection fails to establish property dualism, being based upon a theory of dispositions that is discredited.

(b) *Differences in location*

Place next makes the claim that a disposition and its alleged categorical base cannot be identical because they have different locations. The microstructure is inside the entity, the disposition 'in so far as it is located anywhere, is outside the entity at its point of interaction with other things.'[24] We are given a 'most striking' example of a dispositional property that is located outside the entity which nevertheless possesses it: the magnetic field of an iron bar. Another case is that of opium's dormitive virtue, or *virtus dormitiva*, which manifests itself inside the organism that ingests it.

This argument would be uncompelling to a committed identity theorist, for it seems deniable without absurdity. The disposition, if it is to be located, can simply be located wherever the categorical properties responsible for its manifestations are. Indeed, it seems that the opposite view has the greater absurdity. If the disposition is not to be located in the object itself, then in virtue of what does it belong to that object? One explanation of why Place may think that the disposition is to be located outside the object is that he is susceptible to the mistake of conflating the manifestation of the

[24] *Dispositions: A Debate*, 61.

disposition with the disposition itself. Is the *virtus dormitiva* of opium really to be located at the point where the opium interacts with the organism that has ingested it? The intuition that the opium would still have had the disposition whether or not anyone had ingested it suggests not. This interaction between opium and organism is the event of the disposition manifesting itself; it is not the disposition itself, hence we should not look there for the location of the disposition. What of the most striking example: that of the magnetic field of an iron bar? That something is magnetic means that it has properties that occupy the functional role of attracting certain objects when they enter a certain close proximity. It must be properties within the object that occupy this role but how they do so is another matter. Our understanding is that a field is created that can capture other magnetic objects. If the magnetism is being ascribed to the field, then both the disposition and those properties occupying that role, whatever they are, are to be located within the field. If the magnetism is being ascribed to the object, in virtue of creating a field, then the properties we are interested in are inside the object. In either case, there seems no reason why we are obliged to locate the disposition and its base at different places.

On this issue it is also worth a passing comment on Place's proviso that he is concerned with the location of a disposition *in so far as it is located anywhere*. Place is evidently expressing a degree of scepticism that it makes sense to speak of dispositions having a location. This I take as evidence of further confusion resulting from the conditional analysis of dispositions. Disposition ascriptions are evidently being reduced to conditionals, rather than understood as properties; and certainly it is a category mistake to speak of a conditional, or any other proposition, having a spatial location. Treating dispositions as properties that make conditionals true instantly dissolves the absurdity.

(c) *Differences in causal role*

Place's final argument is that dispositions and their bases occupy different causal roles and so, in virtue of this, cannot be identical. The aetiological structure Place supports is one where a categorical base causes the disposition and the disposition, in turn, causes the manifestation (alternative [A] from Sect. 5.2). Although Place makes these claims in the context of a more fully worked out ontology, they are strange in a number of respects. One may well ask why it is not possible that a categorical property causes the disposition manifestation on its own, directly, instead of creating this separate dispositional property to do the job? Perhaps one response could be that only dispositional properties can cause their manifestations and categorical bases are causally impotent with respect to these.[25] But this is straightforwardly inconsistent with the claim that the categorical base causes the disposition because it means that the categorical property is capable of manifesting a disposition: a disposition to create a further disposition. Even more serious, perhaps, is the concern that this causal structure is an infinitely regressive one. In order to cause a particular disposition manifestation, on this story, a disposition must first be caused by the structure. In order to cause this, though, the structure must have a disposition to cause the disposition that causes the particular manifestation. To manifest this disposition must it first cause a disposition to do so? To do so it must first create another, and so on. It seems, therefore, that a categorical property can have no disposition unless it has an infinity of dispositions, which is a strong reason for thinking this aetiology mistaken.

I will be promoting an alternative but it should be clear that independent consideration of this account of the causal relations between dispositions and their bases shows that it is a highly problematic one. The alternative I promote will be less problematic but it will be one where the categorical base is a cause of the disposition manifestation (alternative [B] from Sect. 5.2). In fact, the categorical base will have the same causal role as the disposition, hence it will be of no use to Place as it directly contradicts the premiss of this argument for dualism.

In chapters to follow I will be considering the causal relations between dispositions, bases, and laws of nature in more depth. For now I will pass on, noting that a convincing argument for the ontological distinctness of a disposition and its base is not to be found in Place's discussion.

[25] See Sect. 5.7: potentiality cannot be reduced to actuality.

5.6 *The Argument from Rigid Designation*

Prior, Pargetter, and Jackson have one more argument to offer against identity and this is perhaps the most serious argument in respect of the damage it does to a monist position of all those discussed so far. The argument from rigid designation originates, in a different context, with Kripke.[26] Using the disposition of fragility as an example, Prior, Pargetter, and Jackson say:

> Accordingly if 'fragility (being fragile) = having α (say)' is true, it is necessarily so, and if false, necessarily so . . . But there are worlds where fragile objects do not have α, for it is contingent as to what the basis of a disposition is. Hence there are worlds where 'fragility = having α' is false for the decisive reason that the extensions of fragility and being α differ in that world; and therefore by rigidity it is false in all worlds, including the actual world.[27]

If we take Dx to represent x having a particular disposition and Cx to represent x having a particular categorical base, then the argument is:

[a] *alleged identity:* $D = C$
but: $\Diamond\exists x\,(Dx \,\&\, \neg Cx)$
therefore: $\Box\neg(D = C)$.

In words: property names rigidly designate; that is, they pick out the same things in all possible worlds, so if a disposition D is identical to a categorical property C it will be identical in all possible worlds. There are, however, possible worlds where there is some particular x that has D but not C. If D is not identical with C in one world, then it is identical with C in no world. Hence necessarily it is not the case that D = C; that is, 'D = C' is false in all worlds, including our own, and even if all instances of D coincide with instances of C in our world.

Of course, Kripke allows that this sort of argument works in some cases but not in others. What is crucial is whether we are picking out the thing in the putative identity relation by its essential or accidental features. If we pick it out just by accidental features, it might be that we can imagine these features distinct from that to which it is identical when such distinctness is not

[26] S. A. Kripke, 'Identity and Necessity', in M. Munitz (ed.), *Identity and Individuation* (New York, 1971).
[27] 'Three Theses about Dispositions', 254.

really a possibility. The example Kripke uses of a genuine (necessary) identity is the identity of heat and molecular motion. Where heat (H) is identical to molecular motion (M):

[1]	*alleged identity:*	$H = M$
	though it seems:	$\Diamond\exists x\ (Hx\ \&\ \neg Mx)$ [and $(\Diamond\exists y\ (\neg Hy\ \&\ My))$]
	but this can be explained:	$\Diamond\exists\varphi$ (φ is the epistemic counterpart of H and $\neg(\varphi = M)$).[28]

The rigid designation argument cannot be used to dispute the identity statement in this case. We identify heat by a contingent feature of it, namely, the sensation it produces in creatures with our peculiar perceptual organs, which of course, could have been different. Our sensations of heat are not, therefore, essential features of heat but are accidental. The objection does not work because it is these sensations that we can imagine occurring without molecular motion, not heat itself; for heat just is molecular motion. Such sensations constitute the epistemic counterpart of heat; that which corresponds to heat in our experience.

However, in the case of the alleged identity of pain (P) and c-fibre firing (C), there is an important difference:

[2]	*alleged identity:*	$P = C$
	though it seems:	$\Diamond\exists x\ (Px\ \&\ \neg Cx)$ [and $(\Diamond\exists y\ (\neg Py\ \&\ Cy))$]
	and there is no epistemic counterpart of pain:	$\neg\Diamond\exists\varphi$ (φ is the epistemic counterpart of P)
	from which it follows:	$\Box\neg(P = C)$.

There is no epistemic counterpart available for pain because feeling painful is an essential rather than an accidental property of pain; that is, anything that feels like pain to us is actually pain as there is no distinction possible between the phenomenon, pain itself, and our experience of the phenomenon. A possible world in which pain exists without c-fibre firing is thus sufficient to defeat the alleged identity of pain and c-fibre firing, even if they coincide completely in the actual world.

The effectiveness of Prior, Pargetter, and Jackson's argument from rigid designation depends, therefore, on whether the alleged identity between a disposition and its causal base is analogous to

[28] Kripke's example needs some qualification: it applies only to convective and not radiative heat.

the tenable identity in [1] or the untenable identity in [2]. Can we say that in imagining a possible world where disposition D was not identical to C we are merely imagining an epistemic counterpart to D: something which is our experience of D but is not D itself? Can we, that is, defend the following argument for dispositions?

[3] *alleged identity:* $D = C$
though it seems: $\Diamond\exists x\,(Dx \,\&\, \neg Cx)$ [and $(\Diamond\exists y\,(\neg Dy \,\&\, Cy))$]
but this can be explained: $\Diamond\exists\varphi$ (φ is the epistemic counterpart of D and $\neg(\varphi = C)$).

I will not attempt to answer this question until Chapter 7. It is worth noting now, however, the relationship between this argument and the earlier argument from variable realization that I allowed had some prima facie plausibility.[29] Is the variable realization of a disposition sufficient to show that the argument from rigid designation must work against the identity theorist? These issues clearly have some connection for the variable realization in fact, of a disposition, illustrates the point that it is indeed contingent what categorical properties base what dispositions. This contingency, and its importance, will be discussed later.

5.7 *Potentiality and Actuality*

Thompson, quoted above (Sect. 5.2), expressed an argument that might be considered the strongest of all for a division of properties into dispositional and categorical. I list the argument here because I think it sets a problem for any anti-dualist account. This is another argument that I think will require a lengthy answer that I will develop over a number of chapters. The argument was this: 'suppose that the exact shape and size of an object were known, the shapes and sizes of all the constituents, along with a list of these facts at every time. We would still know nothing about how or why the object would change with time or on interactions.'[30]

Thompson has set a challenge which the anti-dualist ontology must meet: if dispositions are (type or token) identical with categorical properties such as shape and structure, then how can it be

[29] Sect. 5.4, above. [30] 'Real Dispositions in the Physical World', 69.

possible to know everything about the shape and structure of a thing without thereby knowing anything about the causal powers or possible behaviour of that thing? The dualist position provides one answer to the question: a thing has dispositional properties, which are distinct from any categorical properties, in addition to all its categorical properties. The problem faced by the monist is one of explaining the potentiality of things in terms of what is wholly actual. There must be some reason why objects or properties are able to do some things and not others. A separate class of properties, causal powers, may appear to be a plausible explanation of this but the property monist does not allow that dispositions are different in nature from other properties.

The capacity to answer this problem can be used as a standard for gauging the adequacy of any monist alternative. The monist ontology I develop will have an answer to it. The most expanded versions of the answer will be found in Chapter 10.

5.8 *Problems for Dualism*

I have examined a number of arguments and found varying degrees of effectiveness. Some of the arguments have been dealt with sufficiently but some will have to be assessed in more depth later. We can say, for now, that there is at least some plausibility in the distinctness of dispositions and their categorical bases; hence some plausibility in property dualism for dualism is false if distinctness is. The arguments that I grant initial plausibility to, and will answer as an alternative to dualism emerges, are thus:

(i) Dispositions are variably realized by different categorical bases, hence cannot be identical with any of them (Sect. 5.4).

(ii) Disposition predicates are rigid designators. There are possible worlds where they are not identical with their categorical bases therefore they are identical with their categorical bases in no world, including our own, and even if they coincide completely with those categorical bases in our world (Sect. 5.6).

(iii) Dispositions must be properties extra to categorical properties because a thing's categorical properties do not determine its dispositional properties. Hence, potentiality cannot be reduced to actuality (Sect. 5.7).

Before I pursue the arguments for dualism further however, I want to consider some of the consequences of dualism. We have seen that dualism of dispositions and their bases can be crafted into a plausible position but what would be the status of dispositions if it were true?

My strategy is this: I aim to show that the distinctness of dispositions from their categorical bases is problematic. I show this in order to motivate the search for an alternative to the dualistic ontology, an alternative that does not suffer the same problems and can accommodate our pre-theoretic intuitions about dispositions. Only when I have shown this alternative to be preferable will I return to the arguments for distinctness and give an account which avoids them.

I will mention three problem areas for dualism: parsimony, the causal impotence of dispositions, and the brand of functionalism about dispositions desired by supporters of dualism.

(a) *Parsimony*

Distinctness of a disposition from its base is a less ontologically parsimonious position than that of identity. An identity theory means that there is just one type of property with different ways of talking about that property. It may have to be admitted by some monists that their ontology will also need to draw on a notion of laws to explain why each categorical property has the dispositions it has but a notion of laws will also be required for the dualist's connection of dispositions and their bases.[31] The basic charge against the dualist is that a fully worked out monist theory will be able to explain everything that the dualist theory can explain but with fewer types of property. If this is indeed so, then the monist theory is preferable.

The principle at work here is simply that attributed to Occam: that entities should not be multiplied beyond necessity. This itself need not count against distinctness if such distinctness is unavoidable. The point is, though, that it puts the onus on the dualist to

[31] This is the issue that occupies many of the exchanges in the early chapters of Armstrong, Place, and Martin's debate. ch. 10, below, will give an account of monism without laws.

justify dualism's multiplication of types of property. Parsimony may be considered just good methodology in our ontological constructions, which is certainly not a sufficient reason *per se* for the dismissal of a position, but it does tempt us to the view that the necessity of additional components of an ontology must be justified, rather than a position that declines to add these components at no cost to the explanatory power of that position. If one position can account for everything another accounts for, but does so using fewer entities, then the position using the fewer, other things being equal, is to be preferred.

The dualist response to this needs to show, therefore, that there is something that the monist fails to accommodate. Should it be impossible to find any such thing, and there is no further argument to decide between the two positions, we may decide between the two positions on the basis of parsimony. Certainly we would not be basing our decision on an a priori principle; we would just be applying a cautious methodology. I think, however, that this scenario under-represents the monist's case because there are positive arguments to be offered for monism and there are further difficulties for dualism.

(b) *The impotence of dispositions*

A big problem for the dualist position arises when we consider the causal status of dispositional properties; for here, it appears, the dualist offers a counter-intuitive account.

Prior, Pargetter, and Jackson articulate their position in three theses. The first is the causal thesis that all disposition manifestations are caused and this cause of the disposition manifestations is known as the causal base of the disposition. This thesis I shall not object to at any point in my argument for we would not want to say that it is uncaused what disposition manifestations occur. The second thesis is that of the distinctness of a disposition from its causal base. I have given, above, some of the arguments it is possible to use in support of this position. It is, I maintain, a controversial thesis, but one that has not yet been refuted. However, let us consider the third thesis. This, the causal impotence of dispositions, is a necessary consequence of theses one and two, for if the base is what causes the manifestations, and the disposition is distinct from the base, then the disposition does not cause the manifestations.[32] If dispositions were causally impotent, it is worth noting, then the term 'disposition-manifestation' would have to be deemed a misnomer for this is usually understood as involving a causal relation between the disposition and its so-called manifestation. A disposition, in this case, is allowed no manifestation in this sense. Instead we would have to speak of the manifestations of various categorical properties.

This third thesis, I maintain, casts suspicion on the conjunction of the first two; for it is counter-intuitive in a number of respects that dispositions be causally impotent. One respect in which it is counter-intuitive is that we may wonder in what way a causally impotent property exists at all.[33] Another respect is that dispositions are, as a matter of conceptual necessity, the causal properties of a thing. Take these causal properties away and we no longer have dispositions as normally understood. I take such a thesis of the impotence of dispositions to be so counter-intuitive that it may even be considered a *reductio* of the conjunction of the first two theses. However, not everyone agrees that the impotence of dispositions is counter-intuitive, indeed some may regard this as a virtue of the theory, so I will assume that some argument is necessary to dismiss the position.

(c) *Prior's functionalism*

Prior's favoured position for dispositions is that they are functional properties of things.[34] This, in itself, seems a good way to characterize dispositions. However, when coupled with the ontological dualism of dispositions and their bases, that Prior supports, then the functionalism about dispositions looks less credible.

Prior tries to make room for dispositions in the dualistic account. To have a disposition is to have a functional property; but what is it to have a functional property? It is to have 'a second-order property as opposed to a first-order one'. It is 'the property of having a property that plays a particular causal role'.[35] On the assumption of distinctness, this is problematic. Given that dispositions are causally impotent then we can question what, within the

[32] Provided we accept certain qualifications concerning underdetermination and overdetermination which I examined in my 'Dispositions, Bases, Overdetermination and Identities' and return to in ch. 6, below.

[33] Cf. Sect. 6.2, below.

[34] *Dispositions*, ch. 7.

[35] Ibid. 81.

dualist context, the having of such a functional property adds to its possessor. By the impotence thesis, it adds nothing. All properties that play causal roles, as we have seen, are categorical properties. This raises the awkward question for Prior's functionalism of how a causally impotent property can be functionally characterized. To be functionally characterized is to be characterized according to causal role but dispositions on this theory have no causal role. Hence what it is that is being characterized functionally must be the causally efficacious properties of the particular—the categorical properties—and this starts to look like a monistic theory where a disposition ascription is just the giving of a functional characterization to a categorical property. Here it appears that there is only a division in the way we speak about properties, not a division of properties themselves. In Prior's dualistic functionalist theory, dispositions start to look wholly superfluous; and if they are superfluous, the theory collapses into monism. For this reason, the position has the appearance of being self-defeating.

5.9 *Three Better Theses about Dispositions*

Prior, Pargetter, and Jackson outlined their property dualism in three theses. It should now be clear where the main contrasts between their position and monism are going to be made. Support was given by Prior *et al.* for the theses:

(1) each disposition has a causal base;
(2) each disposition is distinct from its causal base,

which together entailed:

(3) dispositions are causally impotent.

I have argued that, notwithstanding theses (1) and (2), thesis (3) is counter-intuitive. But, as I have said, (1) and (2) entail (3). The *desideratum* of replacing the Prior *et al.* theses with three better theses shapes the approach of the next few chapters. (3) is to be rejected and in the next chapter I will take time to show why arguments to the contrary—that dispositions are causally impotent—are not compelling. If we are to replace (3) with:

(3′) dispositions are causes,

as the ordinary concept of a disposition requires, then at least one of (1) or (2) must be rejected. My choice will be to reject (2), given that the rejection of (1) appears inconceivable. I will replace (2) with:

(2′) each disposition is identical with its causal base,

and (1), (2′), and (3′) form a consistent set.[36] I aim to demonstrate a bit more than consistency, however, so in Chapter 7 I offer arguments for the truth of (2′) and try to defeat, finally, the arguments for dualism. The argument of the next chapter, that dispositions are causes, becomes doubly compelling because, as will be seen in Chapter 7, this forms a premiss in the argument for monism.

[36] This was the basic argument structure of my 'Dispositions, Concepts and Ontologies', in H. Wallis (ed.), *Language and Related Matters* (Leeds, 1994).

6

Dispositions as Causes

6.1 *Causes and Causal Explanations*

Dispositions are properties and properties play causal roles in a thing's interactions with the world about it. This is the justification for my thesis 3′. A more developed justification involves answering the objections to this claim and describing fully the way in which dispositional properties can have a role in bringing about their manifestations.

The issue of the causal role of dispositions is entwined with the question of the use of disposition ascriptions in explanation. Disposition ascriptions have an explanatory value which justifies the use of the dispositional idiom. Their ascription must have explanatory content for there would be no point in making them otherwise. But what is this content? There is little agreement about exactly what explanatory role disposition ascriptions play or even that dispositions have an explanatory role at all.

I will be justifying the claim that disposition terms are a class of explanatory concepts. We have a dispositional vocabulary to use when there is an explanatory gap in our account of how something behaves the way it does. Dispositions fill such gaps, if only temporarily, and in order to fill such gaps for any time at all dispositions cannot be causally impotent nor disposition ascriptions wholly pointless. What I propose to do in this chapter is bolster the causal status of dispositions against objections and clarify the explanatory role of disposition concepts. I begin with some prima facie claims about dispositions which I aim eventually to legitimize.

Prima facie, dispositional explanations are a brand of causal explanation. If we consider the questions of why some objects break when dropped and why some substances dissolve when in water, one kind of answer is that such things break and dissolve because they are fragile and soluble respectively. These statements may be true to the extent that the disposition terms 'fragile' and 'soluble' are true of the objects or substances involved but these statements are also offered as *explanations* of breaking and dropping, and this is the point that is controversial. Being fragile is alleged to be a causally relevant property to the object breaking when dropped; being soluble is alleged to be a causally relevant property to a substance's dissolution upon immersion in water. It needs to be acknowledged straightaway that these explanations are not being offered as complete causal explanations. Such facts about the possession of these properties only explain their manifestations given the right (ideal) background conditions and stimulus events, as described in Chapter 4.

The justification of the causal potency of dispositional properties can be set out in the following argument. Not all substances will dissolve upon immersion in a specific volume of water in a certain set of background conditions. Some will dissolve and some will remain in their solid state. We could have two substances in identical test conditions and one shows a certain reaction while the other does not. This scenario is best explained by the theory that something about the substance is relevant to the behaviour it exhibits, otherwise, given that all other conditions are the same, everything would exhibit the same behaviour in the same circumstances. It is manifestly false that everything behaves the same for any test and set of background conditions. It seems that we are obliged to grant some causal role to the particular object or kind involved. It is this causally relevant *something* about the object that is the property of the object that can be called, according to an acceptable use of 'cause', the cause of the manifestation. If this property can be described correctly as a dispositional property, then we have a justification for the thesis that dispositions are causes.

I shall refer to this argument as *the argument from behavioural difference*. A more precise statement of the argument would be as follows:

1. For some test F and set of background conditions C_i, there exists at least one x and one y for which reaction G is true of x and false of y.

2. This difference in behaviour of x and y is best explained by the possession by x (or y) of some causally relevant property or property complex P, not possessed by y (or x).

3. There are circumstances in which this property P can be correctly described as a dispositional property.

Therefore: dispositions are causally relevant properties.

Premiss 1 is an empirical premiss which I take to be unproblematically true and confirmable with ease. Premiss 2 rests on a theoretical claim about the cause of a thing's behaviour. It relies on the claim that if x and y differ in behaviour, then there are cases where this difference in behaviour has a cause; that is, there are cases where the behaviour of x and y is non-indeterministic. If all other factors are the same, then the best explanation of such difference in behaviour is a difference in properties. The possibility of this case is not undermined by the possibility of another case. This is where x and y differ in behaviour but are identical in all their properties. The coherence of this rests on the possibility of x and y possessing the same probabilistic disposition which just happens to get manifested for x but not for y. The argument is not affected by this admittedly plausible case for it merely requires the equally credible possibility of the first type of case. In premiss 3, I claim that there are circumstances where these properties can correctly be said to be dispositional. This depends on a semantic and ontological theory for dispositions. The semantic theory has been introduced in Chapter 4; it will be completed in Chapter 9. The ontological theory will be introduced in Chapters 7 and 8. Already enough has been said to indicate how it is possible to correctly refer to such properties in a dispositional way: if property P is characterized functionally, then P is being characterized as a disposition. This leads to the conclusion that in the sense of causal relevance stated in premiss 2, dispositions are causally relevant.

How can the argument from behavioural difference be challenged? It has been suggested, by the opponents of realism about dispositions, that dispositions are valueless as causal explanations, basically because dispositions are not causes of anything. A disposition ascription, on this view, is akin to a *virtus dormitiva*: a trivially true ascription of a causal power that we posit when we are in ignorance of the 'real' causes of changes in things. A *virtus dormitiva*-type explanation is allegedly a vacuous explanation in that it offers an explanation for how it is that something has a power to φ by ascribing that very same power to one of its components. Such explanations are thought to be highly undesirable and it is hoped, for instance by Quine, that eventually, as science progresses, *virtutes dormitivæ* will be completely eliminated from explanation and be replaced with 'respectable' categorical causal explanations.[1]

What I will do in the rest of this chapter to support the causal efficacy of dispositions is first, in Sect. 6.2, show why it would be undesirable to argue that dispositions are causally impotent. I then tackle a number of objections to the causally efficacious view of dispositions. These are that dispositions are of the wrong ontological category to be causes (Sect. 6.3), that appeals to dispositions are redundant in scientific explanation (Sect. 6.4), and that dispositional explanations are trivial explanations (Sect. 6.5). The line of attack culminates in the most famous objection: the *virtus dormitiva* objection (Sect. 6.6). I show how dispositional explanation can be non-trivially true (Sect. 6.7) though also that there are cases where such triviality must be conceded (Sect. 6.8). This is not to be regarded as a weakness for dispositions, however. It is a consequence of the meaning of disposition ascriptions. They are property terms which already contain an indication of causing particular manifestations in particular circumstances. Dispositions are understood as causes *par excellence* (Sect. 6.9). The champion of the causal efficacy of dispositions thus has no need to defend dispositions against a common triviality charge. Dispositions as causes becomes a sustainable claim.

6.2 *Causal Efficacy and the Existence of Properties*

The upshot of the objections is that dispositions are not causes of their manifestations, indeed not causes of anything; they are wholly causally impotent. This conclusion is significant. We saw in Chapter 1 that the most natural ontological category in which to place dispositions is that of properties. The objections against dispositions as causes cast doubt on the classification of dispositions as genuine properties. The doubt cast is a result of the implications of causal impotence on any putative property's existential status. If the objection is conclusive and dispositions are causally impotent, then

[1] *Roots of Reference*, 8–15.

their existence as real properties is threatened by the adoption of the following criterion for the existence of non-abstract properties:

The causal criterion of property existence: for any intrinsic non-abstract property *P*, *P* exists if and only if there are circumstances *C* in which the instantiations of *P* have causal consequences,

which was the principle implicitly assumed in the worries I expressed in Sect. 5.8(b) about Prior's dualistic ontology.

The causal criterion requires explanation. First, this is a criterion which applies only in the case of intrinsic, non-abstract properties. By 'intrinsic' I mean non-relational properties or monadic universals. It is arguable that relational properties, if they are properties at all, need bestow no causal powers on the particular in which they are instantiated. There are cases, for instance, where there is a change in a relational property of a particular; this is a so-called 'Cambridge change' such as a change from being one thousand metres from a burning barn to being one thousand and one metres from a burning barn. A criterion of property existence need not rule out such relational properties on the basis of their possession bestowing no causal power; hence the restriction that it applies only to monadic universals. The criterion need not apply to abstract properties either, such as being divisible by 2, for which no causal potency is claimed.

The causal criterion applies to the *instantiations* of a property *P*. Some realists about universals may argue that the universal itself is something that exists over and above its instantiations. A universal of this kind, existing not in the physical realm, would be wholly causally inert. My formulation of the causal criterion deals with the instantiations of properties because these are things which should be expected to be involved in causal interactions if the property in question is a genuine one. If the putative instantiations of *P* have no causal consequences, then the *P*-predicate is to be judged an empty term with no reference. However, such causal interactions of the instantiations of *P* need not be ongoing all the time for *P* to qualify as a property. Only in certain circumstances are they brought into action. What is crucial is that there are some possible circumstances in which the possession of *P* makes some difference to the particulars that possess it.

What is the argument for the acceptance of such a causal criterion? The criterion rules out the existence of any property that has no possible causal effects but why? Couldn't it be that there was a property that had no possible causal consequences in any circumstances in any of its possible instantiations but nevertheless was really possessed by some object? There are two ways in which we could attempt to justify the criterion. One way would be to offer a developed theory of universals in which a necessary connection was postulated between properties and causal powers. Perhaps this could be developed to the point where non-abstract, intrinsic properties are viewed as *being* causal powers. I think that this way of understanding properties is likely to be the most profitable and, if true, it renders the causal criterion of property existence true analytically.[2] I have been implicitly working with this kind of account of properties and will be making further claims that support it. I will attempt first, however, to justify the criterion in another way which is not dependent on the adoption of this theory of universals. I will, however, go on to show how this line of thought, supported by Armstrong and others, suggests the theory of universals I would favour.

According to the *virtus dormitiva* objection there could be no possible time at which a disposition could issue forth an effect, for the argument suggests that dispositions are causally impotent necessarily: they are logically precluded from causing their manifestations because of their analytic connection with such events. Thus there could be no difference in the way the world was if one minute a particular gained a disposition, then lost it the next minute, then gained it the next, and so on. When we say that something has gained a property we seem, for this reason, committed to the thesis that its causal powers have been altered in some way. On the analysis of dispositions that would seem to follow from the objections, however, it would make no difference whether a particular had ten dispositional properties, ten thousand, or none.

This supports a theory of universals which differs from Armstrong's but is nevertheless supported by much of what he says. He speaks, for instance, of 'the link which it is natural to make between the properties of things and the causal powers of things.'[3]

[2] For discussion of some of these issues see E. Fales, *Causation and Universals*, xiii. Fales endorses something similar to the causal criterion: 'a physical universal exists if and only if it is a member of the causal "web": if it is then it exists whether or not it is ever instantiated.'

[3] D. M. Armstrong, *A Theory of Universals* (Cambridge, 1978), 20.

There is a historical precedent for this kind of view in Plato. In the *Sophist,* the Stranger says:

I suggest that anything has real being that is so constituted as to possess any sort of power either to affect anything else or to be affected, in however small a degree, by the most insignificant agent, though it be only once.[4]

Armstrong outlines further this 'link between universals and *causality*'. He proposes a number of principles which include:

(i) that every property bestows some . . . power upon the particulars of which it is a property;

(ii) that a property bestows the very same power upon any particular of which it is a property, and

(iii) each different property bestows a different power upon the particulars of which it is a property.[5]

Further support for this kind of view comes from Shoemaker who says: 'what makes a property the property it is, what determines its identity, is its potential for contributing to the causal powers of the things that have it.'[6]

The support from these sources for the causal criterion is qualified, however. Armstrong denies that such a principle can be supported as an a priori truth. The justification for all three of his principles is, he says, no more than 'pragmatic'; but he thinks that this is justification enough:

[I]t seems possible to conceive of a property of a thing which bestows neither active nor passive power of any sort. But if there are such properties, then we can have absolutely no reason to suspect their existence. For it is only in so far as properties bestow powers that they can be detected by the sensory apparatus or other mental faculty.[7]

Such properties would be logical possibilities, on Armstrong's view, but we would have no reason to think them actual. Shoemaker says that the reason to believe such a criterion is 'epistemological': only with such an understanding of properties can we explain how they are capable of 'engaging our knowledge'.[8]

[4] 247d-e. [5] *A Theory of Universals*, 42–3.
[6] 'Causality and Properties', 212. [7] *A Theory of Universals*, 44–5.
[8] 'Causality and Properties', 214.

Clearly, much hangs on the prior question of universals in general. If properties are causal powers, a causal criterion of property existence does follow a priori.[9] There is no necessity for purposes of the present debate of pursuing this issue to a definite conclusion, however. It suffices to say here that if it can be shown that dispositions in principle have no causal effects, then the claim that they are real properties is under threat and the question of the legitimacy of a causal criterion of property existence would have to be engaged fully. It seems, therefore, that there are a number of alternatives now open. The main ones are:

1. Acceptance of the causal impotence of dispositions and acceptance of the causal criterion of property existence. This option would justify eliminativism about dispositions and some form of categorical monism; that is, the claim that all genuinely existing properties are categorical properties. Eliminativism about dispositions is the claim that there are no properties that disposition terms refer to; hence reductionism about dispositions, which posits identity relations between dispositional and categorical properties, is to be rejected.

2. Acceptance of the causal impotence of dispositions but rejection of a causal criterion of property existence. This option would not seem to be an easy one because it would necessitate giving an account of properties where there need be no connections between a thing's properties and its causal powers. I have little optimism for any such account. I will, instead, be opting for a third alternative; namely:

3. Rejection of the causal impotence of dispositions and acceptance of a causal criterion of property existence. This option is also no easy task because of the multiple objections to the potency of dispositions. As well as the *virtus dormitiva* objection there will be other objections to face: that dispositions are states, and states are not causes, and that complete 'scientific' explanations require no reference to unexplained powers. I will be considering each of these objections.

[9] The Stranger in the *Sophist* seems to want something like this also; for he goes on to say at the same place: 'I am proposing as a mark to distinguish real things that they are nothing but power.'

6.3 *States as Causes*

Cummins gives the following argument:

Dispositions are states, hence not causes, whether we take these states to be irreducibly dispositional in character (brute) or identify them with some non-dispositionally specifiable state (a 'categorical basis')[10] . . . To attribute a disposition *d* to an object *a* is to say that *a* satisfies a set of conditions such that any of a certain kind of events E would constitute a completing condition (cause) for a manifestation of *d* (effect).[11]

The objection is that dispositions are of the wrong ontological category to be causes: they are states, hence not events, and the only thing that can be a cause is an event. I am not sure that Cummins intends such states to differ in any significant way from properties, or at least complexes of properties. I will assume, therefore, that as properties are not events, Cummins's argument would apply equally to properties as it does to states. Note that this argument is also applicable to categorical properties; hence the Prior position, where a disposition has a categorical causal basis is also under attack. In supporting the view that dispositions can be causes I am thus against both the Prior and Cummins lines of attack.

In a simple sense, Cummins's argument is correct. A disposition is a property and a property may be thought of as something which on its own does nothing. For a disposition manifestation to be caused it is necessary, for the most part, that there be a change; specifically, an initiating stimulus, which is an event and can, therefore, be a cause of the manifestation. The possession of such a disposition is neither a change nor an event and can, therefore, initiate nothing.

However, it is clear, I would argue, that an acceptable sense of 'cause' can apply to dispositions, for what can be said to constitute a cause may encompass many different kinds of ontological item which, here, Cummins is limiting to something akin to 'initiating cause' or Aristotle's 'efficient cause'. It is not claimed that the causal efficacy of dispositions is causally efficacious in the sense of an efficient cause. Nevertheless, some causal role can be attributed to

[10] Cummins's own footnote: 'acquisition of a disposition is an event, and hence can be a cause.'

[11] R. Cummins, 'Dispositions, States and Causes', *Analysis*, 34 (1974), 200.

a property. The aetiology of the manifestation event for an ordinary stimulus-manifestation disposition can be described as follows: a stimulating event *c* occurs which, together with and only together with a property *d* of *x*, causes a 'disposition manifestation' response event *e* in 'ideal conditions' C_i.

There may be a number of different types of event that can result in *e*, given *d*, just as the exact categorical nature of *d* is left unspecified by a dispositional characterization of *d*. All that matters for the dispositional characterization of *d* is that it is a property that will result in *e*, given *c*. Thus *c* and *d* are causes of *e* only in conjunction. Both are necessary conditions for *e*, in that both *c* without *d*, and *d* without *c*, would result in the failure of the manifestation of *e*. Thus an object that is fragile has a property that in ideal conditions, as explicated in the previous chapter, can cause breakage when there is a suitable stimulus event, such as the object being dropped; and for any object that is dropped, this event can cause breakage only if the object has a property that is causally apt for breakage in such circumstances, namely, the object is fragile. This means that *c* and *d* are jointly necessary and sufficient for *e*, if the disposition is a deterministic one, or jointly necessary and sufficient for the determination of the probability of *e* where the disposition is probabilistic.

To speak of a disposition as the cause of a manifestation *e* must, therefore, be understood with the proviso that it is the cause of *e* conditional upon the antecedent stimulus; or, *d* causes *e* only if in conjunction with *c*. Thus there is a change of the sort Cummins considers to be necessary for a disposition manifestation; a change in the conjunction that causes, or is the initiating stimulus of, an effect; even though one of the conjuncts—the disposition—may remain static throughout this change.[12]

To speak of causes is always to tell only part of the story: a cause of an event is never the whole cause of that event. Given that causal chains may stretch indefinitely into the past, and involve

[12] Consistently with a principle of reciprocity, it seems reasonable to say that when a dispositional property becomes manifest it also undergoes change e.g. when a sample of sugar manifests its solubility in dissolving it undergoes alteration in molecular structure which results in the solubility being lost, i.e. sugar in solution can no longer be said to be soluble. Similarly, an explosive substance may not be explosive after it has been exploded. Would this kind of consideration be another way in which Cummins could be answered? It seems that we can say that a disposition manifestation involves some change involving the dispositional property.

indefinitely many factors, when we ask for the cause of an event we may selectively look at what, at a particular time and place, is identifiable as a causal antecedent, or part of the causal history, of the *explanandum* event. In the case of dispositions we are concerned with that part of the immediate causal history which is constituted by what it is about the object or substance in question, stated in terms of its properties, that is a causal antecedent of its issuing the manifestation event *e*. It is not claimed that this cause in the object or substance is causally sufficient for *e*: it requires other background laws and conditions, including an antecedent stimulus.

It seems, therefore, that there is no decisive threat to the causal efficacy of dispositions in the fact that they are, in Cummins's opinion, states; nor if they are construed, as I put it, as properties or qualities.

6.4 *Do Explanations Appeal to Dispositions?*

The paradigm of scientific explanation, it is alleged, is structural explanation that contains no reference to dispositional properties. Quine argues in support of such an interpretation of explanatory practice as dealing only in the categorical.[13] He concedes that there remains some remaining reference to dispositions in scientific explanations but thinks that this merely indicates that physical science is still incomplete. As it becomes more advanced such references to dispositions will be eliminated in favour of a completed explanation that contains appeal to categorical properties only. The contention is that we can, in principle if not yet in practice, provide a full causal explanation of all the events that are 'pre-scientifically' understood as disposition manifestations in terms that are entirely 'disposition-free'. We can offer a description of the categorical bases that are relevant to the occurrence of the putative disposition manifestation event and these properties can replace the unscientific appeal to dispositional powers or forces.[14]

However, reference to certain categorical properties clearly falls

[13] *Roots of Reference*, 11–12.

[14] This is one interpretation of Quine's claims. In ch. 8, I consider an alternative interpretation.

short of an explanation if, following Hume, we allow no necessary connections between such categorical properties and the events they may precipitate. Something more is needed if we are to have anything that is a candidate for a complete explanation. One view, with essentially Humean origins, is that it is necessary to subsume the behaviour resulting from such properties under general laws, thus correlating kinds of properties with kinds of events.[15] O'Shaughnessy does this in his explanation of 'elevancy', being the disposition of a particular object to rise when in water. A scientific understanding of this phenomenon will not be in terms of objects having a real disposition or power, says O'Shaughnessy, but rather in terms of the categorical properties of the 'elevant' particular and the general laws connecting such properties with certain types of behaviour. Thus O'Shaughnessy explains the rising of an object in water in terms of the satisfaction of five conditions:

(1) The ratio of the object's mass to its nonpermeable volume is less than one.
(2) The object has weight.
(3) The object tends to move in the direction of the impressed force.
(4) The density of water is 1.
(5) Water exerts an upward thrust on objects equal to the weight of the water they displace.[16]

What it is about the object that causes it to rise in water is its so-called 'basis', which on this view is non-dispositional: '*The basis* of a disposition I take to be that categorical state of affairs in the "disposed" that determines the presence of its disposition; and this is identical with the causal contribution of the "disposed" to the manifestation.'[17] All he need now argue is that a disposition and its causal base are distinct and the disposition is thereby rendered impotent—powerless, as he puts it—because there is no causal role left for it to play. O'Shaughnessy sees bases as distinct from dispositions because bases explain dispositions, thus 'the base is not the disposition it helps to explain, and therefore such dispositions are powerless.'[18]

It is intended, of course, that this type of disposition-free explanation for elevancy be generalized. We may account for the ability

[15] What Cummins, in 'Dispositions, States and Causes', calls the *subsumption strategy*.
[16] 'The Powerlessness of Dispositions', 2–3.
[17] Ibid. 10.
[18] Ibid. 5.

of a billiard ball to roll when struck in terms of it being an equally balanced sphere on a flat surface. We may account for brittleness in terms of weak bondings between the constituent molecules. We may account for the explosiveness of dynamite in terms of its chemical composition. All explanations, it is claimed, will be non-dispositional explanations.

McMullin similarly offers a defence of disposition-free explanation. He sets out a defence of what he calls *structural explanation* where a structure is defined, somewhat vaguely, as 'a set of constituent entities or processes and the relationships between them'.[19] Despite obviously failing to do so, it is clear from the rest of McMullin's paper that this definition is meant to exclude dispositions. A *nomothetic explanation* is the alternative to a structural explanation. Nomothetic explanations are appeals to laws where the *explanandum* is shown to be an instance of a general law. Such explanations will be essentially *deductive-nomological* in form and Hempel indeed argues that dispositional explanations are a species of deductive-nomological explanation.[20] For example, that a glass is fragile and that it is struck are jointly sufficient for the glass to break and the sufficiency resides in the nomological connection between being fragile, being struck, and breakage. The weakness of such nomothetic explanations, McMullin maintains, is that they appeal to laws that regulate behaviour without explaining the cause behind that regularity. Structural explanation postulates a mechanism; 'it is an opening up of a hitherto hidden world of processes and structures both macroscopic and microscopic.'[21]

The search for mechanisms to explain powers is, of course, no novel project: it was Boyle's chief concern in a number of places.[22] McMullin admits that a mechanism is for the most part not known, but *postulated* in our explanations, hence such explanations are *hypothetico-structural* or HS explanations.[23] The virtue of H-S explanation is that it is the explaining of an effect in terms of a cause rather than in terms of it being an instance of a regularity: it seeks to explain why a disposition was manifested, not state that its manifestation was an instance of a regularity. It is argued, therefore, that structural explanations are more informa-

[19] 'Structural Explanation', 139.
[20] 'Dispositional Explanation.'
[21] 'Structural Explanation', 145.
[22] e.g. his *Origin of Forms and Qualities.*
[23] 'Structural Explanation', 139.

tive explanations, which are thus genuine explanations, superior to trivial, uninformative dispositional explanations, and are what we should aim at in presenting scientific explanations.

The explanatory success of the H-S model is qualified, however. Consider McMullin's explanation of the power of aspirin: '[t]he operation of aspirin is not yet well understood but one guess is that it has to do with the shape of the aspirin molecule and its ability to "lock into" certain parts of the brain.'[24] There is a clear appeal to a disposition here: an ability to 'lock into'. This cannot be dismissed as a mere oversight—an inadvertent lapse—on McMullin's part. Rather, I would argue, it is an inescapable corollary of such explanations, and is evident also in O'Shaughnessy's account of elevancy. O'Shaughnessy's condition (3) appealed to a *tendency* to move in a certain direction, condition (5) appealed to water *exerting* an upward thrust, and condition (2) was that the object had weight, which could be construed as a dispositional property. The problem for such an account is that a world described entirely in the terms of the explanatory units of McMullin's H-S scheme—structures, shapes, and microstructures—is a world without change that stands in need of animation. For the events, processes, and movements of the world to be explained fully we must allude to what such structures can and will do: how they will behave in a variety of situations, what processes they can be involved in, and how they will be affected by other objects and substances. Some reference to how something behaves can never be eliminated from the account if we are to aim for complete explanation. Again, what is missing is something which connects classes of properties with classes of events. Thus McMullin's structural explanation unintentionally reduces to (a refined) nomothetic explanation. A disposition manifestation is explained in terms of the possession of a structure but this explanation is incomplete. To become complete the explanation must invoke a law or the possession of a dispositional property such that the possession of that structural property will result in a certain manifestation. Something must explain why a particular, no matter what its structural properties, will ψ if φ-ed. We may appeal, therefore, to a structure S, but we may still have to invoke a general

[24] Ibid. 145.

law or dispositional property connecting S to ψ. The explanation, on McMullin's account, will inevitably slide into the form:[25]

$$Sx$$
$$\forall x\ ((Sx\ \&\ \varphi x)\ C \rightarrow \psi x)$$
$$\text{Therefore, } \varphi x\ C \rightarrow \psi x.$$[26]

Therefore, if nomothetic explanations are trivial, as McMullin suggests, H-S explanations suffer from the same defect and are really no alternative at all. McMullin's attempt to avoid the 'triviality' of the nomothetic explanation is rather like the attempt to refute the Humean thesis, that any effect may follow from any cause, by appeal to some intermediate mechanism that necessitates the effect. The same problem emerges when Montuschi suggests 'structure' as '"a ground" for explaining the occurrence of aspects and observable behaviours of entities'. This supposedly 'performs an ontological transformation of the concept of cause' from a Humean constant conjunction to a real existent.[27] The point is that the same principle applies to this more sophisticated cause, from which any effect may follow. What connects *this* structure to the *explanandum* event? McMullin and Montuschi appeal to a structure to explain a regularity but it so explains only if it appeals to a further regularity;[28] so as an exercise in avoiding appeals to regularities, it fails.[29]

What does this say of scientific explanation and the claimed dispensability of appeals to dispositions? There is no evidence that our explanatory practice dispenses with nomothetic explanation or dispositional explanations. Arguably our scientific explanations depend essentially on appeals to dispositions and we should not ignore the fact that certain entities, which are fundamental to

[25] I will be arguing in chs. 9 and 10 that this Humean derived form of explanation is not our only alternative.

[26] My conditional 'C→' is a stronger-than-material conditional for expression of a realist commitment to causal relations.

[27] 'From Effects to Causes', 25f.

[28] Or permits the existence of real dispositional properties: the option I will come to endorse.

[29] To be fair, Montuschi does allow a dispositional aspect to structures, which may mean that the line of criticism I offer is escaped.

A similar argument to those I have discussed is advanced by R. Squires, 'Are Dispositions Causes?', *Analysis*, 29 (1969) against Armstrong's 'reduction' of dispositions to categorical bases in *A Materialist Theory of the Mind*; this strategy 'succeeds' in explaining only if it ascribes another disposition to the categorical base and if this disposition is similarly reduced an infinite regress is initiated.

modern physics, can be characterized only dispositionally.[30] The behaviour of subatomic particles cannot be further analysed into structures and this may tempt us to regard these as instances of 'brute' ungrounded dispositions which end any possible regress of explanation. Any claim that scientific practice, as a matter of fact, has dispensed with nomothetic or dispositional explanation is simply false. The Quinean response would be that these so-called 'brute' dispositions will be eliminated as we make our physical theory complete but, if the analysis in this section is correct, a more advanced theory would still have to appeal to further ungrounded or unexplained regularities.

I will be going on to give an alternative interpretation of scientific explanation and practice to that of the H-S model in Chapter 10. I will suggest that science aims at the discovery of general truths which are grounded in the dispositions of classes of object and that in the case of the basic constituents of the physical world, these dispositions are 'ultimate' insofar as there is no further explanation, categorical or otherwise, as to why these particulars have the dispositions they have.

What I have tried to do in this section is to show that causal explanation in terms of categorical properties alone is inadequate. This realization, it is hoped, will show the motivation for finding some 'active' powers in the world. This section has also gone some way to illuminating the explanatory role of appeal to dispositions, as connections between an object's states and its behaviour.

6.5 *The First Triviality Objection*

What are we looking for in a causal explanation? David Lewis says that 'to explain an event is to provide some information about its causal history.'[31] This requirement seems easy to satisfy. That a particular had a disposition to ψ prior to it ψ-ing would seem to qualify as part of the causal history of it ψ-ing, for Lewis's statement is neutral as to what sort of entities are parts of a causal history. As he says: 'information about what the causal history includes may

[30] For a more detailed account of appeal to forces in modern physics see P. Davies, *Superforce* (Penguin, 1995) which contains details on the current disposition-laden accounts of matter.

[31] D. Lewis, 'Causal Explanation', in *Philosophical Papers*, ii (Oxford, 1986), 217.

range from the very specific to the very abstract.'[32] What is debatable is whether reference to dispositions identifies part of a *causal* history. There are various ways we can causally explain badly:[33] we can give false information, we can give little information, or we can give information that was already possessed. It is the latter charge that the triviality objections identify as the reason that dispositional explanations are no good as causal explanations.

There are two objections that disposition ascriptions are trivial causal explanations and hence not causes of their manifestations. The second is the *virtus dormitiva* objection, which I come to in the next section, but this differs from the first type of triviality objection that I describe now.

The first triviality objection to dispositions having a causal role, and hence having any role in causal explanation, goes as follows. To be fragile means nothing more than to break if dropped; to be soluble means nothing more than to dissolve if immersed in water. A similar analysis can be given for any disposition: to say that something possesses a disposition is just to say that the appropriate response will follow upon the appropriate stimulus. But if solubility means just 'dissolves in water', then any explanation of why a substance dissolves in water in terms of it being soluble will be nothing more than a trivially analytic explanation, which is no explanation at all. If fragile means nothing more than 'breaks when dropped', then it is no explanation of why something breaks when dropped.

This argument depends on the premiss that nothing can be the cause of something to which it is logically connected. So if dispositions are logically connected to their manifestations, they cannot be causes of those manifestations. Mackie puts this point in Humean terms: 'specifically dispositional properties . . . would violate the principle that there can be no logical connections between distinct existences, which . . . is the least disputable step in Hume's critical discussion of causal necessity';[34] that is, if the cause must be logically independent of the effect, then dispositions cannot be causes of their manifestations because, given certain antecedent conditions, a disposition would logically entail its manifestation, whereas all genuine causal relations are logically contingent rather than necessary.

[32] D. Lewis, 'Causal Explanation' in *Philosophical Papers*, ii (Oxford, 1986). 220.
[33] Summarized by Lewis, ibid. 226–7.
[34] 'Dispositions, Grounds and Causes', 104.

This objection relies on a certain understanding of dispositions. We must, that is, construe a disposition ascription along Rylean lines, namely:

[Df_R] x is $D =_{df}$ if x is F-ed, then x will G.

If we accept Df_R, then we cannot explain why x G-ed, when F, in terms of x being D because x being D means nothing more than that x will G, if F.

It can be acknowledged that there is some initial credibility to Df_R; for it is, indeed, in virtue of something inducing sleep when ingested that we call it dormitive, it is in virtue of something breaking when dropped that we call it fragile, and in virtue of something dissolving when in water that we call it soluble. Whatever arrangements of structures or processes are at work in such manifestations are irrelevant to the meaning or truth-conditions of these ascriptions. For dispositions in general, it is purely in virtue of the satisfaction of such functional criteria that they are the dispositions they are. This explains the analytic connection between a disposition and its manifestation: 'solubility' has a conceptual connection to 'dissolves in water in ideal conditions', so if in water, soluble and in ideal conditions, dissolving is entailed analytically.

But dispositions are not really subjects of quite so much triviality and hence can be causally efficacious, for the Rylean analysis of disposition ascriptions is incomplete. I argue for functionalism, rather than Rylean behaviourism, about dispositions. As I showed with the argument from behavioural difference, there must be some reason why not everything breaks when dropped from a moderate height. There must be something about the subject of a disposition ascription which makes the ascription true—something which plays a causal role in dissolution in water and breakage when dropped. This, and only this, must be added to Df_R:[35]

[Df_M] x is $D =_{df}$ x has some property P (and P is a cause of x G-ing if x is F-ed in conditions C_i).[36]

[35] As I argue in Sect. 4.5, nothing further about the nature of this causally relevant property can be known solely from an analysis of meaning.

[36] E. Sober, 'Dispositions and Conditionals, or, Dormative Virtues are no Laughing Matter', *Philosophical Review*, 91 (1982), 594, gives a similar definition and thus agrees with me, against Ryle, that disposition ascriptions are not simply equivalent to subjunctive conditionals.

This puts a stop to immediate triviality, for it is no longer trivial to explain why a substance dissolves in water by saying that one of its properties is responsible.

Some doubts that may remain about the triviality of dispositional explanations could, however, resurface in a slightly different way. This is the second triviality objection that disposition ascriptions are *virtutes dormitivae.*

6.6 Virtus Dormitiva

The objection that dispositions constitute *virtutes dormitivae* explanations is endorsed by Armstrong and Mackie.[37] This objection is one that differs from the plain triviality objection though the two are easy to confuse.

The *virtus dormitiva* label comes from Molière's joke at the expense of philosophers. When asked why opium puts one to sleep when ingested, the candidate philosopher answers that it does so because it has a *virtus dormitiva*—a dormitive virtue:

Bachelierus: I am asked by the learned doctor for the cause and reason that opium makes one sleep.
To this I reply that there is in it a dormitive virtue,
whose nature is to make the senses drowsy.

Chorus: Very, very, very, well answered. The worthy candidate deserves to join our learned body.[38]

[37] Armstrong, *Belief, Truth and Knowledge*, 16 and Mackie, 'Dispositions, Grounds and Causes', 104.

[38] 'Bachelierus: Mihi a docto Doctore
Domandatur causam et rationem quare
Opium facit dormire:
A quoi respondeo,
Quia est in eo
Virtus dormitiva,
Cujus est natura
Sensus assoupire.
Chorus: Bene, bene, bene, bene, respondere.
Dignus, dignus est entrare
In nostro docto corpore.'

J.-B. Molière, *La Malade Imaginaire*, in *The Plays of Molière*, trans. by A. R. Waller (Edinburgh, 1926), 328. The English trans. is a slight variation from that of K. Hutchison, 'Dormitive Virtues, Scholastic Qualities, and the New Philosophies', *History of Science*, 29 (1991), 245.

The charge of triviality here differs significantly from the first case in that it is not a charge the force of which rests upon assuming the Rylean account. It cannot, therefore, be dismissed on those grounds. Some causal power in the opium seems to be acknowledged by Bachelierus. He is asked what in the opium causes sleep when ingested and this question, and his answer, in acknowledging a cause of sleep in the opium, is acknowledging something that the Rylean analysis does not. The triviality of the exchange can be reduced to the paraphrase:

Q: what property in the opium causes it to induce sleep upon ingestion?
A: a dormitive virtue; that is, something that causes sleep upon ingestion.

This clearly does not rely on Rylean assumptions because the dormitive virtue in question is consistent with a functionalist understanding of dispositions as:

[Df_M] x is $D =_{\mathrm{df}} x$ has some property P (and P is a cause of x G-ing if x is F-ed in conditions C_i).

This brings into focus the question of whether disposition ascriptions are akin to *virtutes dormitivæ* ascriptions. My answer is that they are, but in what follows I justify their use despite the usual objections to *virtutes dormitivae.*

What objection to dispositions as causes can be built out of this second charge of triviality? In the first place there is the charge brought that a disposition cannot cause its manifestation because it is logically connected to it. This is a problem that will have to be dealt with but there is also a deeper concern that the *virtus dormitiva* case points to. The suggestion seems to be that dispositional properties cannot be properties at all for they are just inventions of the sophist who is ignorant of the true causes of events. Given any event, and the assumption that every event has a cause, then some power can always be invented as the cause of that event. But if that power ascription just means a cause of such an event then the power ascription will be trivially true and uninformative.

The questions to be faced now are whether to concede the triviality charge for dispositions and, if the charge is conceded, whether the triviality is one fatal to the causal status of dispositional properties.

I suggest that there are two kinds of question that can be asked about the cause of a disposition manifestation and that whether a power ascription is trivial is different in each case. The first kind of question is exemplified in:

Q1. Why, whenever opium is taken, does sleep follow?

Here, a *virtus dormitiva* explanation can be true non-trivially, as in the case of 'uncombable hair syndrome', below. However, there is a second type of question exemplified in:

Q2. Why does opium make one sleep?

and here a *virtus dormitiva* explanation is true trivially. Nevertheless, I argue that, in spite of this, dispositions can still be causes of their manifestations. I will deal with each type of question in more depth.

6.7 *'Uncombable Hair Syndrome'*

It has been argued that understanding a disposition ascription as implicitly carrying a claim of a causally relevant property is all that is required to defeat the objection: if we construe dispositions as functional properties, then it tells us that the cause of sleeping was in the opium and this is to rule out various other possibilities. It rules out opium merchants administering a genuine soporific each time they see someone take opium,[39] it rules out sleep being a constant but accidental conjunct with the ingestion of opium, it rules out divine intervention.

Hutchison considers the apparent *virtus dormitiva* in the statement:

> Researchers in the United States have confirmed what some parents have long suspected—that you cannot do a thing with some children's hair. And they believe unruly hair is caused by a little-known condition called uncombable hair syndrome.[40]

Hutchison takes uncombable hair syndrome to be akin to a *virtus dormitiva* but thinks it not useless as an explanation of unruly hair, for it assigns a responsibility for the unruly hair to the hair itself. This rules out parental self-delusion about their child's hair, atmos-

[39] D. Lewis, 'Causal Explanation', 221.

[40] 'Dormitive Virtues, Scholastic Qualities, and the New Philosophies', 246–7.

pheric disturbance moving the hair, other children having ruffled it, a psychosomatic cause, the laziness of the child when it comes to combing, and witchcraft. Hutchison holds that *virtutes dormitivae* explanations may be false, hence they are not trivial. There were, in Molière's time, a number of putative explanations that competed with that of the scholastic qualities—the powers—that were his target. Some alternatives are: empiricism, that denies any such causal efficacy in the opium; reductionism, that reduces *virtutes dormitivae* down to fundamental, underlying, causal powers; fictionalism, an instrumentalism about powers; occasionalism and supernaturalism.

However, Hutchison's strategy is not satisfactory. He is answering questions of type 1, above, instead of questions of type 2. This type of answer would have sufficed only if Bachelierus had been asked why, whenever opium was taken, sleep followed. If this had been the question, the response that sleep was caused by a property of the opium would have been informative. Bachelierus was asked why opium *makes* one sleep, which already acknowledges the causal responsibility of the opium; thus witchcraft and atmospheric disturbance are already ruled out in the terms of the question. In answer to the question of why something causes a certain effect, the charge of triviality must be allowed to stand.

6.8 *The Cause of* G *Caused* G

A better response is to concede the triviality of *virtutes dormitivae* and dispositional explanations whenever offered as an answer to type-2 questions. After all, on the view I think the most convincing, dispositions are causal-role occupying properties by definition. They are functionally defined properties and a functionally defined property is bound to be trivially connected to its functional, in this case causal, role. This point can be conceded but the triviality of a disposition ascription's explanatory content is not lethal to the causal efficacy of the disposition ascribed. Compare Df_M, above, with δ:

[δ]: The cause of *G* caused *G*.

δ would be an inadequate causal explanation of *G* but we do not want to say that, because of this, the cause of *G* was not the cause of *G*, for that would be a contradiction. Similarly, if by a disposition term *D*, we mean the cause of *G*-ing upon being *F*-ed, as I claim we do mean by a disposition term, then it would be a nonsense to claim that the cause of *G*-ing upon being *F*-ed was not the cause of *G*-ing upon being *F*-ed because it is logically connected to *G*-ing upon being *F*-ed.

This is little more than an endorsement of the division between the extensionality of the causal relation and the intensionality of the relation of causal explanation. If A causes B, it does so no matter how A is described (and no matter how B is described); but for description α of A to be part of an informative causal explanation of B, under description β, the intensions or meanings of α and β are relevant in that they fix their referents by picking them out in a certain way, by a certain mode of presentation. Whether these modes of presentation can be related in an informative way depends on what they are. To dismiss dispositions from causal efficacy on the ground that they provide bad causal explanations is thus to put the cart before the horse. Weakness of a causal explanation does not preclude causation between events though without causation, there is no true causal explanation holding between those events. Nothing about ontology is at stake in questions of explanation for explanatory success is contingent upon the modes of presentation of *explanans* and *explananda* and relative states of knowledge and ignorance.[41]

It is worth, at this point, noting the parallel between this problem and its proposed solution and Davidson's solution to the problem of whether reasons are causes.[42] Intentions are analytically connected to actions hence, one argument goes, they cannot cause those actions because to do so would violate Hume's principle of the contingency of connections between causally related events. Davidson's solution is to concede that intentions provide no causal explanations of actions but they nevertheless can be causes of them. Whether one event causally explains another is dependent upon the modes of presentation of the two events and this permits no mental causal explanation of actions. The causal

[41] I return to this theme in Sect. 10.6.

[42] D. Davidson, 'Actions, Reasons and Causes', in *Essays on Actions and Events* (Oxford, 1980).

relation is an extensional relation, however, thus it is blind to the distinct mental and physical modes of presenting events. Therefore, a mental event can cause a physical event though it provides no causal explanation of it. In the case of dispositions we can say, analogously, that a functionally characterized property is so characterized because it causes events of a certain kind and this entails that it has a conceptual, necessary connection with such events. As such it cannot suffice as a causal explanation of such events but it may be possible to redescribe it in a way that is not analytically so related. The causal relation, being extensional, holds between such properties and such events no matter how we refer to them; hence it is blind to the dispositional–categorical distinction. Therefore, we can say that dispositions cause their manifestations even though they do not causally explain them. Given the correct understanding of dispositional properties, the denial that they are causally efficacious of their manifestations becomes as absurd as the denial of δ (the cause of *G* caused *G*) on the ground that the cause of *G* is analytically related to the cause of *G*. Certainly δ is a trivial causal explanation of *G* and thus cannot be counted as an explanation at all for, while a trivial truth is still a truth, a trivial explanation is not an explanation. For something to explain it must be non-trivial.

6.9 *Causes* par excellence

The functionalist theory treats concrete, non-abstract, dispositions as causes *par excellence*. The example of 'the cause of *G*' in δ was more than just an example of an uninformative attempt at causal explanation. The characterization of a property dispositionally is the description of a property according to (at least one of) its functional role(s); hence it is the reference to a property in a manner like 'the cause of *G*' rather than in a manner like 'molecular composition H_2SO_4'. Dispositions can thus regain the metaphysical role traditionally ascribed to real powers: the that-in-virtue-of-which-something-will-*G*, if *F*.

This illustrates how the functionalism about dispositions I advocate differs from that of Prior, which was shown to be problematic. I see dispositions as causal-role occupying properties, by definition. Curiously, Prior agrees that dispositions are functional properties; that is, presumably, causal-role occupying properties.[43] But she also asserts that dispositions are causally impotent: they cause nothing.[44] How a causally impotent property can be classified according to its causal role is beyond the understanding of the present author and is certainly not a position he recommends. Indeed, it is puzzling that McLaughlin cites Prior's functionalism as 'the leading theory of dispositions today'.[45]

Because dispositions are causes *par excellence*, they can, in certain contexts, have an informative and explanatory role. Their ascriptions can be informative when they indicate what potencies lie within the objects of ascription. They indicate what events will be causes in what conditions within an assumed context. Their ascriptions can be explanatory. A certain occurrence can be attributed to a property of an object which causes the occurrence, rather than to some external condition such as atmospheric disturbance. That two objects differ in their behaviour can be explained in terms of their having different properties rather than there being other conditions active upon them. Hence, while disposition ascriptions can have a trivial, uninformative role in certain contexts, they can have non-trivial and informative roles in others.

6.10 *A Return to Ontology*

The objections to causal efficacy as an ontological thesis have been sufficiently discredited and an analysis of dispositions has been offered that describes dispositions as causal-role occupying properties by conceptual necessity. On this view, dispositions are causes though they can be poor causal explanations of their manifestations. The poverty of such explanations resides in the fact that they provide no detail of the mechanisms involved in a disposition manifestation, they tell us only that a property is possessed which plays such-and-such a causal role, thus the property is characterized only functionally, not 'structurally'. This necessary causal-role occupancy is thus the source of the *virtus dormitiva* objection, but

[43] *Dispositions*, ch. 7.

[44] 'Three Theses about Dispositions' with Jackson and Pargetter.

[45] B. P. McLaughlin, 'Disposition', in J. Kim and E. Sosa (eds.), *A Companion to Metaphysics* (Oxford, 1995), 123.

it is an objection that is avoided if we deem it as having application only to causal explanation, rather than causation itself.

The objections have been disarmed and an analysis of disposition ascription has been offered but something more is needed, namely, a plausible ontology in which there is a causal role for dispositions to play. We see, with the Prior, Pargetter, and Jackson position, an ontology that leaves no room for dispositions to have a causal role. What has been offered so far cannot be taken as conclusive because, as has been shown, this question appears to be one that is contingent upon prior ontological questions. This message also comes out of the debate held in the pages of *Analysis*.[46] Squires claims that if Armstrong is right, then categorical bases do all the causal work leaving no causal role for dispositions. Armstrong responds that the disposition and its categorical base are identical, therefore dispositions are causes.

What I have done in this chapter is show that the objections that dispositions are a priori causally impotent are not secure. This means that dispositions are not logically precluded from being causes and so an ontology cannot be dismissed simply because it involves a causal role for dispositions. It is to such ontological issues, therefore, that I now return.

[46] R. Squires, 'Are Dispositions Causes?' and 'Are Dispositions Lost Causes?', *Analysis*, 31 (1970); L. Stevenson, 'Are Dispositions Causes?', *Analysis*, 29 (1969); D. Coder, 'Some Misconceptions about Dispositions', *Analysis*, 29 (1969); D. M. Armstrong, 'Dispositions are Causes', *Analysis*, 30 (1969); O'Shaughnessy, 'The Powerlessness of Dispositions' and Cummins, 'Dispositions, States and Causes'.

7

Property Monism

7.1 *Monism Replaces Dualism*

Having considered the question of whether dispositions can be causes, I return now to the point in the argument I left at the end of Chapter 5. I noted in that chapter that there were problems for any dualist theory of the relation between dispositional and categorical properties. These were problems of parsimony, of the causal role played by dispositions and of a coherent formulation of functionalism about dispositions. I showed at the end of the chapter that if we wanted to say that dispositions were not causally impotent, then this required the rejection of at least one other of the theses supported by Prior, Pargetter, and Jackson: (i) that each disposition has a causal base or (ii) that each disposition is distinct from its causal base. It is the latter thesis—the distinctness thesis—that I choose to reject. I replace this thesis with its contradiction: dispositions are not distinct from their causal bases, they are identical with them in a sense to be explained. The three new theses—the causal thesis, the identity thesis, and the potency of dispositions—form a consistent set.

My job in this chapter is to formulate and support an identity theory for dispositions and their so-called categorical bases. In particular, I must describe and defend a version of property monism and then show how the arguments for distinctness that are outstanding from Chapter 5 are no threat to the kind of property monism I recommend. I hope also to improve upon my earlier statement of monism[1] and specify more explicitly some of the corollaries of the theory. Having looked at the nature of the identities involved, I will then consider the property monist's response to two atypical examples of putative dispositions: abstract dispositions and ungrounded dispositions. These are cases where it might be thought property monism cannot apply. I will concede that

[1] 'Dispositions, Bases, Overdetermination and Identities'.

there exists some doubt as to whether a theory of dispositions ought to attempt to accommodate these cases but as I think the right theory of dispositions can accommodate them there is at this stage no harm in including them in our considerations.

One of my chief aims will be to develop a clear articulation of the nature of the monist claim and that is where I will begin.

7.2 *The Monist Claim*

In Chapter 4, I argued in support of a conceptual distinction between the dispositional and the categorical. I showed how we could distinguish between these two forms of predication. The thesis of property dualism tries to take this distinction further, into ontology. Property monism, on the other hand, is the claim that the conceptual distinction is the *only* distinction because it denies that there is an ontological division in reality which the conceptual division maps. The world contains instantiated properties. There are not distinct categorical and dispositional types of properties but there are distinct categorical and dispositional ways of talking about instantiated properties. What is crucial to property monism, therefore, is a justification for the claim that dispositions and their categorical bases are actually the same states of particular things, characterized in two different ways, rather than distinct states in the world.

The chief argument for identity I offer is an argument from identity of causal role where, roughly, the numerically identical causal roles of any two tokens, p_1 and p_2, entails the identity of p_1 and p_2. These two tokens could be a disposition and its base where two predicates pick out states, or instantiations of properties, of an object that make exactly the same causal contribution to the behaviour of that object. From this assumption, the argument directs us towards the conclusion that there can only be one state involved which the two predicates pick out in their different ways.

This argument gains support, though not always in the explicit case of dispositions, from Lewis, Peacocke, and Shoemaker.[2] Alston is also supporting this view when he says:

[2] D. Lewis, 'An Argument for the Identity Theory', *Journal of Philosophy*, 63 (1966), and 'Psychophysical and Theoretical Identifications', *Australasian Journal of Philosophy*, 50 (1972), 249; C. Peacocke, *Holistic Explanation*, 134f.; and S. Shoemaker, 'Causality and Properties', 232–3.

My candidate [for non-synonymous predicate identity] is the principle of causal relevance. According to this principle, two-state designations refer to different states of a substance if and only if the states referred to have different bearings on the causal interactions into which the substance might enter. If the states referred to make exactly the same contribution to any[3] causal interactions into which the substance might enter, then we should regard them as one and the same state under different descriptions.[4]

The argument that is suggested can be given a more formal expression as:

The argument from identity of causal role
1. disposition d_1 = the occupant of causal role R
2. categorical base c_1 = the occupant of causal role R
Therefore: disposition d_1 = categorical base c_1,

where 'the occupant of causal role R' is intended as a definite description and 'categorical' is understood in the sense described in Chapter 4.[5] In the next section I will be attempting to justify the argument but first I want to get clear what the conclusion means: what it means to say that a disposition and its categorical base are identical.

The first step in understanding the intended identity of dispositions and categorical bases is the rejection of what Armstrong calls the Argument from Meaning.[6] The Argument from Meaning would have it that any two non-synonymous property predicates designate different properties merely in virtue of the fact that they are non-synonymous predicates. The denial of the Argument from Meaning is the claim that the same property can be designated by two or more terms that are neither phonetic-orthographically, nor semantically, identical; that is, the terms are neither tokens of the same word nor tokens of two words that mean exactly the same.[7] This means that the identity conditions for properties are separated from the identity conditions for predicates. As Armstrong

[3] Alston leaves an undesirable ambiguity here, as I will show in Sect. 7.7. It is desirable that he mean '*every* causal interaction', rather than '*a single* causal interaction'.

[4] W. P. Alston, 'Dispositions, Occurrences, and Ontology', in R. Tuomela (ed.), *Dispositions* (Dordrecht, 1978), 374.

[5] This argument structure is adapted from Lewis, 'Psychophysical and Theoretical Identifications', 249.

[6] D. M. Armstrong, *A Theory of Universals*, 103.

[7] Armstrong, ibid. 7.

puts it, this gives us the 'emancipation of the theory of universals from the theory of semantics'.[8]

This separation of ontology from meaning creates the possibility of an identity theory of non-synonymous property predicates, of which the mooted identity theory of dispositions and their categorical bases is an instance. In supporting such a theory we may make use, in a different context, of the famous Fregean distinction between *Sinn* and *Bedeutung*.[9] There, Frege had shown that the same particular, for example Venus, could be referred to by terms which differed in sense, for example 'the morning star' and 'the evening star', such that it was possible to affirm a proposition in which one of these terms appeared though deny a proposition in which a co-referring term was substituted. Whereas Frege's example was of two particulars referred to by non-coextensive definite descriptions, property monism *apparently* requires the reference to a single property by two or more non-coextensive predicates. Some may regard this application of the sense–reference distinction to properties to be objectionable, though its application in the case of a particular is not. I aim to show how, for a variety of reasons, the distinction can hold for the case of dispositions and bases, though not for the same reason Armstrong gives.

Immediately, the sense–reference distinction frees the argument from any objection along the lines of referential opacity; for instance, the meaning of disposition term d_1 can be understood when the meaning of the categorical property term c_1 is not understood so it may be argued that d_1 is not identical with c_1. Such objections can be dismissed as intensional fallacies. It is essential that such arguments can be dismissed because the sort of identities that are likely to be mooted will be theoretic a posteriori identities. This means that some empirical work will be needed for the discovery of true identity statements. Hence I can know what it is for x to be soluble though I need not know what it is for x to have the particular molecular structure which is a categorical base of solubility, say *xyz*. I need not know which molecular structure is involved or even what a molecular structure is, even though 'x is

[8] Ibid. 6.

[9] G. Frege, 'On Sense and Meaning', in P. Geach and M. Black (eds.), *Translations from the Philosophical Writings of Gottlob Frege* (Totowa, NJ, 1980).

soluble' may be true in virtue of the possession of the same state that makes '*x* has molecular structure *xyz*' true.

The identity theory makes the claim, therefore, that each disposition is numerically identical to a causal base that can also be designated in categorical terms. For each state responsible for an event that is a disposition manifestation there are dispositional and categorical modes of designation and these modes can differ in sense when they coincide in their reference.[10] In Chapter 4, I specified how these two distinct ways of describing reality differed. The candidate categorical bases to be identified with dispositions will be states of an object or substance such as its shape, macrostructure, and microstructure.

A number of problems remain outstanding, such as the question of whether it is property types or property tokens that are identified and whether every disposition must have a categorical base. I will put these questions aside until I have developed further the argument for identity.

7.3 *The Occupants of Causal Role R*

I will justify the claim that the argument from identity of causal role is a sound argument. In this section I will offer a justification for the truth of the premisses of the argument. In the next section I will show that the argument is valid.

First premiss: disposition d_1 = the occupant of causal role R

In my earlier presentation of the argument I justified the first premiss by saying that dispositions are causal-role occupiers by definition and that it is a conceptual necessity that, for instance, 'solubility' names the property that is the occupier of the role of causing dissolution on immersion in water.[11] This has now been added to. In Chapter 4, I introduced a functionalist theory which I will discuss more in Chapter 9. A functionalist theory of disposi-

[10] There is the special case of dispositions which lack a categorical mode of designation. See Sect. 7.10.

[11] 'Dispositions, Bases, Overdetermination and Identities', 50.

tions provides the explanation of why it is a conceptual truth that dispositions occupy certain causal roles.

How could the opponent of the functionalist account be persuaded? One line of argument is to point out the conceptual absurdity of dispositions occupying inappropriate causal roles. If it was claimed that something dissolved in water because of its *fragility*, then, unless some explanation could be produced, this claim would appear not just false but also absurd. It would be necessarily false because of the conceptual connection between the causal role of causing dissolving when in water and solubility and the conceptual connection between fragility and the causal role of causing breakage when dropped. The idea of a disposition occupying a different causal role to the one it actually occupies involves a conceptual, rather than a factual, confusion.

I would take it that for any categorical base it is contingent what causal role it occupies, hence there is no conceptual or logical absurdity in a categorical base occupying a different causal role to the one it occupies actually. Such a possibility is only contingently false. This difference may be used as one way of distinguishing the dispositional from the categorical.

Second premiss: categorical base c_1 = the occupant of causal role R

There is a long history of categorical properties or mechanisms being offered as explanations of disposition manifestations. Their explanatory value lies in the empirical or theoretical claims that they occupy, as a matter of fact, those same causal roles which are 'pre-scientifically' understood only in terms of dispositions. Often such non-dispositional explanations are taken as paradigms of scientific explanation that enable us to dispense with the dispositional vocabulary. Robert Boyle provides us with an early example of the 'Mechanical Hypothesis' which seems unreconstructed in contemporary philosophers such as Quine, Armstrong, and Putnam. In Sect. 1.3 I gave some examples from Boyle's works where he provides mechanical explanations for the effects previously attributed to powers, or in the terminology of his day, *qualities.*[12]

[12] The doctrine of qualities is described and discussed by Hutchison, 'Dormitive Virtues, Scholastic Qualities, and the New Philosphies'.

Boyle's view is clearly that such mechanical explanations can be found, in principle if not yet in practice, for all supposed qualities. Certainly, commonplace examples are readily available; for instance, explaining the power of a watch's hands to rotate in terms of the arrangement of springs, cogs, and wheels inside the watch. Boyle thinks that such explanations are to be found even where the powers involve mechanisms too small to be observed. It would be a mistake to attribute mysterious powers on the basis of unknown or unobservable mechanisms. He says:

> [T]here is another sort of philosophers that, observing the great efficacy of the bigness and shape and situation and motion and connexion in engines, are willing to allow that those Mechanical principles may have a great stroke in the operations of bodies of a sensible bulk and manifest mechanism, and therefore may be usefully employed in accounting for the effects and phenomena of such bodies, they yet will not admit that these principles can be applied to the hidden transactions that pass among the minute particles of bodies, and therefore think it necessary to refer these to what they call *nature*, *substantial form*, *real qualities*, and the like unmechanical principles and agents.[13]

This is clearly a close relation of views expressed by Quine, for example, when he says:

> Each disposition, in my view, is a physical state or mechanism. A name for a specific disposition, e.g. solubility in water, deserves its place in the vocabulary of scientific theory as a name of a particular state or mechanism. In some cases, as in the case nowadays of solubility in water, we understand the physical details and are able to set them forth explicitly in terms of the arrangement and interaction of small bodies. Such a formulation, once achieved, can thenceforward even take the place of the old disposition term, or stand as its new definition.[14]

We will see that there are a number of different possible responses to the thesis that categorical properties occupy the appropriate causal roles for dispositions. Boyle's response, for instance, seems to be an eliminativist one. I will discuss these

[13] 'About the Excellency and Grounds of the Mechanical Hypothesis', 142.

[14] *Roots of Reference*, 11. Quine states his views on dispositions at various places: *From a Logical Point of View* (Cambridge, Mass., 1953), 158–9; *Word and Object*, 222–6; *The Ways of Paradox*, 71–4; *Ontological Relativity and Other Essays* (New York, 1969), 130–8; *Philosphy of Logic* (Englewood Cliffs, NJ, 1970), 21; *Roots of Reference*, 8–15; 'Facts of the Matter', in R. W. Shanan and C. V. Swoyer (eds.), *Essays on the Philosophy of W. V. O. Quine* (Hassocks, 1979), 162.

responses at greater length in the next chapter. Now I will accept the thesis as unproblematic and use it as the second premiss in the argument from identity of causal role. The opponents of the argument, if they accept this second premiss, must either seek some way to reject the first premiss, though I have tried to show above that we should not do so, or they must deny that the conclusion follows from these premisses. I look now, therefore, at the validity of the argument.

7.4 *The Validity of the Argument*

The form exemplified by the argument is evidently that of (a):

(a) x = the R
y = the R
Therefore, by transitivity of identity, $x = y$.

This is a valid argument form. If the premisses are accepted, the only quarrel the opponent can have with the argument from identity of causal role is on the ground that it does not in fact exemplify (a). The most credible case for it not exemplifying (a) is that 'the occupant of causal role R' does not refer: it is not a definite description as intended by the term 'the R' in the valid argument form (a).

'The occupant of causal role R' can fail to refer in these two situations: (i) there is no single occupant of causal role R; (ii) there is more than one occupant of causal role R. The first situation occurs where no single state occupies the causal role R, though there may be two things which *taken together* occupy R. This would represent a case of the underdetermination of the disposition manifestation by each of the disposition and categorical base individually. In such a case, a particular would require both a dispositional property and a categorical base in order to exhibit certain behaviour, hence the occupant of causal role R is a conjunction of two distinct properties or states. The second situation would represent a case of overdetermination; for instance where both disposition and categorical base are sufficient for causal role R. Each suffices on its own for the production of the manifestation event. Disposition and base are not identical, though, because they overdetermine the manifestation event.

I have argued that the underdetermination of the manifestation event by the dispositional and categorical properties goes against the intended meaning of the premisses of the argument from identity of causal role.[15] The premisses make the claims, with good evidence, that dispositional and categorical properties are each individually sufficient to facilitate the production of the manifestation event, given the correct antecedent stimulus and background conditions. It is a conceptual claim that dispositions are sufficient for this role and an empirical claim that categorical properties are. However, the possibility of overdetermination is perhaps not so easy to dismiss without ruling it out merely by definition.[16] However, I have argued that the alleged overdetermination would have to be a special type that I have called Peacocke-overdetermination which, I maintain, is an a priori impossibility. I call this type of overdetermination Peacocke-overdetermination because of the formal features it has in common with a case Peacocke discusses. Such overdetermination can be distinguished as follows. 'Ordinary' overdetermination occurs when two or more 'ordinary overdeterminers' occur. Ordinary overdeterminers are properties/states of affairs/events which are individually sufficient (abbreviated as IS) for their effect but are not individually necessary (¬IN) insofar as the effect would still occur had one or more ordinary overdeterminer been absent, as long as at least one ordinary overdeterminer is present. Peacocke-overdetermination consists in the occurrence of two or more Peacocke-overdeterminers. Properties/states of affairs/events are Peacocke-overdeterminers where they are individually sufficient (IS) for their effect but are also individually necessary (IN); that is, the absence of any Peacocke-overdeterminer guarantees the absence of the effect, which is something that ordinary overdetermination cannot guarantee. Peacocke-overdetermination is thus impossible because of the contradiction of two properties/states of affairs/events, *a* and *b*, both being IS and IN. That *a* is IS entails that *b* is ¬IN, and that *b* is IS entails that *a* is ¬IN, *contra hypothesis*. Further, that *b* is

[15] 'Dispositions, Bases, Overdetermination and Identities', 53–4.

[16] Some may declare the possibility of overdetermination to involve a contradiction and hence represent no possibility at all.

IN entails that *a* is ¬IS, and that *a* is IN entails that *b* is ¬IS, *contra hypothesis*. This licenses the following inferences:

(1) $\forall x\,(\mathrm{IS}x \rightarrow (\neg\exists y\,(\mathrm{IN}y\ \&\ (x \neq y))))$
(2) $\forall x\,(\mathrm{IN}x \rightarrow (\neg\exists y\,(\mathrm{IS}y\ \&\ (x \neq y))))$

That is, if any *x* is individually sufficient (or necessary) for an event *e*, then nothing else is individually necessary (or sufficient) for *e* unless it is identical with *x*.

How are we to justify the claim that disposition *d* and categorical base *c* are Peacocke-overdeterminers? In the case where a pain event and physical event overdetermine the withdrawal of a hand, Peacocke says the following: 'Overdetermination implies that even if the pain event had not have occurred, the withdrawal of the hand would still have occurred . . . This we certainly ordinarily take to be false, and it is not clear why we should change the belief.'[17] In the case of the overdetermination of disposition manifestations, the disposition *d* and the categorical base *c* are Peacocke-overdeterminers because *d* and *c* are both causally sufficient for the manifestation event but both are also causally necessary. The disposition manifestation would not have occurred had either *d* or *c* been absent. The warrants for such claims differ. As we have seen, the warrant in the case of *d*, for IS and IN, derives from conceptual necessity; in the case of *c* it derives from an empirical physical theory.

My claims thus far can be summarized. The argument for the identity of the dispositional and the categorical is an argument from the identity of causal role of a disposition and its putative categorical base. For the argument to be sound we must make certain implicit assumptions explicit, namely, that the disposition and its putative categorical base are individually causally necessary and sufficient for the production of a disposition manifestation under appropriate conditions. In such a case the overdetermination of the disposition manifestation is, unlike ordinary cases of overdetermination, an a priori impossibility. Therefore, the causal role R cannot have two such occupiers and the premisses of the argument, that both the dispositional and categorical instantiated properties occupy the causal role R, are tenable only if the dispositional and categorical instantiated properties are identified.

A positive argument for dispositions and categorical bases being

[17] *Holistic Explanation*, 135.

identical, and hence property monism, has been advanced but there were a number of arguments for distinctness and hence property dualism that were said to have some credibility. In the next sections I will treat those arguments for distinctness as objections to identity and consider whether they have sufficient strength to make us abandon the monist position.

I listed a number of arguments in Chapter 5 but judged that only some of them were serious. I judged that the variable realization argument ought to be taken seriously, or at least there ought to be some explanation why, if the claim of variable realization was good, property dualism was not necessitated. The 'swamping' argument I deemed question begging to the extent that it assumed an inappropriate candidate categorical base for a disposition and I said that there was nothing more in the argument than a claim of variable realization. I also allowed that the argument from rigid designation had at least some credibility and I put off a full response until this chapter. I dismissed a number of the other arguments, however; namely, the argument from explanatory asymmetry, the argument from differences of category, the argument from differences of location, and the argument from differences in causal role. It is interesting to note now, however, that this final argument assumes something consistent with the argument from identity of causal role: that difference of causal role entails distinct properties.

The two arguments that can now be answered are the argument from variable realization and, first, the argument from rigid designation. The argument from irreducibility of potentiality, which I also allowed had force, will require a more involved discussion in Chapter 10.

7.5 *Epistemic Counterparts and World Relative Ascriptions*

Prior, Pargetter, and Jackson said:

> [I]f 'fragility (being fragile) = having α (say)' is true, it is necessarily so, and if false, necessarily so . . . . But there are worlds where fragile objects do not have α, for it is contingent as to what the basis of a disposition is. Hence there are worlds where 'fragility = having α' is false for the decisive reason that the extensions of fragility and being α differ in that world; and therefore by rigidity it is false in all worlds, including the actual world.[18]

[18] 'Three Theses about Dispositions', 254.

We saw (Sect. 5.6) that Kripke thought that in some cases this kind of argument from rigid designation worked, namely, where there was no epistemic counterpart to the phenomenon in the identity statement. That is, where:

[2] *alleged identity:* $P = C$
though it seems: $\Diamond\exists x\,(Px\ \&\ \neg Cx)$ [and $(\Diamond\exists y\,(\neg Py\ \&\ Cy))$]
and there is no epistemic counterpart of pain: $\neg\Diamond\exists\varphi$ (φ is the epistemic counterpart of P)
from which it follows: $\Box\neg(P = C)$

in the case of the alleged identity statement 'pain = c-fibre firing'. In other cases the argument did not work, e.g. where it seems that heat could occur without molecular motion. In such a case it was an unreliable intuition that heat could occur without molecular motion and what was being imagined was the epistemic counterpart of heat without molecular motion. Where heat (H) is (necessarily) identical to molecular motion (M):

[1] *alleged identity:* $H = M$
though it seems: $\Diamond\exists x\,(Hx\ \&\ \neg Mx)$ [and $(\Diamond\exists y\,(\neg Hy\ \&\ My))$]
but this can be explained: $\Diamond\exists\varphi$ (φ is the epistemic counterpart of H and $\neg(\varphi = M)$).

In Sect. 5.6 I left open the question of whether the alleged identity between a disposition and its categorical or causal base was analogous to the tenable identity statement in [1] or the untenable identity statement in [2].

An epistemic counterpart in the case of dispositions would not be difficult to find. It could be that our intuitions that a disposition may occur without its alleged categorical base are unreliable because what we succeed in imagining without the categorical base, in some other world, is actually the manifestation of a different disposition, not the disposition in question. I think this possibility is serious because of the evident world relativity of disposition ascription which makes it problematic to draw conclusions from appeal to transworld disposition ascription.[19]

It seems eminently persuasive that our disposition ascriptions are made relative to actual world conditions (or some other set of

[19] See my 'Ellis and Lierse on Dispositional Essentialism', *Australasian Journal of Philosophy*, 73 (1995), 607–8 for the first statement of world relativity of disposition ascription.

conditions fixed by the context of the ascription). In saying that something is fragile, for instance, I am saying that it is fragile in this world; I am saying nothing about how it would behave in some other world which perhaps does not even have the same laws of nature as this world. We can call such an ascribed disposition 'fragility$_{(a)}$', thereby indexing the ascription of fragility to actual world conditions. Take x to have the disposition D_a and take D_a to be the property whose functional role is specified by the stimulus and response pair $<\varphi_a, \psi_a>$, for instance, fragility-in-the-actual-world is the property of being such that, in ideal conditions, if dropped-in-the-actual-world-or-world-with-the-same-laws, then breakage-in-the-actual-world-or-world-with-the-same-laws is caused. In some other possible world $_w$, with different laws, the property which causes breakage when dropped need not be of the same categorical type as in the case of D_a. Hence D_a may be identical to categorical property C_1 but D_w, the property whose functional role is specified by the stimulus and response pair $<\varphi_w, \psi_w>$, is not identical to C_1. This is no problem because it is not the case that $(D_w = D_a)$, given that $<\varphi_w, \psi_w> \neq <\varphi_a, \psi_a>$.

Note that this account does not bind a property to a single world. It may be true of a categorical property C_1 that it has the roles D_{1a} and D_{2w}. If this is true, then it is true in both $_a$ and $_w$, thus $C_1 = D_{1a}$ and $C_1 = D_{2w}$. However, if $C' = D'_a$ and $C'' = D''_w$, and $\neg(D'_a = D''_w)$, then $\neg(C' = C'')$.

This allows the possibility of an epistemic counterpart for D_a because, while it may be true that x is D_a, there is no contradiction in it also being $\neg D_w$ (where $\neg D_w$ is the property such that $<\varphi_w, \neg\psi_w>$). Therefore, we can allow, without contradiction, the set:

$$\{D_a(x), \varphi_w(x), \neg\psi_w(x)\}.$$

This analysis allows us to construct the following solution to the argument from rigid designation:

[3] *alleged identity:* $D_a = C$
though it seems: $\Diamond\exists x\,(D_a x \;\&\; \neg Cx)$
[and $(\Diamond\exists y\,(\neg D_a y \;\&\; Cy))$]
but this can be explained: $\Diamond\exists\varphi$ (φ is the epistemic counterpart of D_a and $\neg(\varphi = C)$),

where φ is filled by $\neg\psi_w$; that is, the manifestation of some disposition that is not identical to D_a.

I think that the world relativized account of disposition ascription is correct and that it could be used in such a way as to defeat the argument from rigid designation were it not for other considerations. The notion of the world relativity of disposition ascription will be called upon again to solve another problem and there will be more of an attempt to justify the analysis. However, in respect of the case against property monism, there remains an argument whose force I will grant: the argument from variable realization. The implication of granting variable realization is that the argument from rigid designation would also go through; that is, in fact, there is a case where:

[3′] *alleged identity:* $D_a = C$
but it is the case that: $\Diamond\exists x\,(D_a x \,\&\, \neg Cx)$ [and $(\Diamond\exists y\,(\neg D_a y \,\&\, Cy))$]
from which it follows: $\Box\neg(D_a = C)$.

In this case, while no doubt there still is an epistemic counterpart for D_a, namely $\neg\psi_w$, it is not this counterpart that is being imagined when we imagine $(\Diamond\exists y\,(\neg D_a y \,\&\, Cy))$ or $(\Diamond\exists x\,(D_a x \,\&\, \neg Cx))$. We do succeed in imagining a case where $\neg(D_a = C)$.

However, what I will show in the next section is that the possibility of variable realization does not threaten the kind of property monism that it is sensible to defend and if the variable realization argument does not threaten it, nor does the argument from rigid designation.

7.6 *Variable Realization and Instantiations of Properties*

We saw that the variable realization argument, which originates in the philosophy of mind,[20] was turned against any putative identity theory of the dispositional and the categorical by Prior, Pargetter, and Jackson. Their argument was:

[T]here is the consideration that it is empirically plausible that certain dispositions have different causal bases in different objects. Suppose in particular that the causal basis of being fragile in some objects is molecular bonding α, in others it is crystalline structure β. (It does not matter

[20] H. Putnam, 'The Mental Life of Some Machines', in *Mind, Language and Reality: Philosophical Papers*, ii (Cambridge, 1975), 415 and 420 being the earliest statements.

whether this is plausible for being fragile, because it can hardly be denied that this happens with some dispositions. And in the case of the disposition of being fit in Evolutionary theory there are a multitude of different bases.)

We cannot say both that being fragile = having molecular bonding α, and that being fragile = having crystalline structure β; because by transitivity we would be led to the manifestly false conclusion that having molecular bonding α = having crystalline structure β.[21]

One possible reponse for the monist is a denial that such variable realization occurs but, in spite of one attempt towards this conclusion,[22] this line is a difficult one to follow and it need not be pursued if variable realization and a form of property monism can be reconciled.

I suggest that they can be reconciled in something like the following way. The property monist essentially wants to show

[21] 'Three Theses about Dispositions', 253.

[22] In my 'Dispositions, Supervenience and Reduction'. The basic line to pursue, if one does want to argue the point, is as follows. There are two mistakes made in drawing a conclusion of variable realization. The first is the taking of conditions for the identity of categorical properties that are far stronger than the conditions for identity of dispositional properties. Thus, according to the identity theory, if *a* and *b* are alike in their possession of disposition D_1, they should be alike in their possession of a categorical property C_1. But according to the variable realization argument, *a* and *b* need not be alike in their categorical properties though they are alike in D_1; therefore $D_1 \neq C_1$. The sceptical issue I raise is whether *a* and *b* really are alike in possession of D_1. Could it be that the property term D_1 is a far vaguer one than those with which we are classifying categorical properties? To use Armstrong's favoured terminology, could it be that D_1 is a determinable whereas the categorical properties in which *a* and *b* differ are determinates? Perhaps *a* and *b* differ in their determinate dispositional properties also, thus providing no proof of variable realization.

The second mistake, that of the inclusion of causally irrelevant properties in the identity statement, is a line of attack on variable realization that I tried to develop in 'Dispositions, Supervenience and Reduction', 430–2. Though I shall attempt to give no great detail here, this response is basically that if all categorical properties that are of no causal relevance to the manifestation event are excluded, as they admittedly ought to be, then there is no guarantee that type–type identity statements will not be the result.

Together, these responses suggest that there is a certain arbitrariness, and hence, unimportance in drawing the distinction between type–type and token–token identities because Armstrong's analysis of the determinable–determinate distinction (*A Theory of Universals*, 111–13) seems to imply that for any two particular tokens *a* and *b*, there will always be some level of description at which they are of identical types and there will also be some level of description at which they are of different types. Variable realization thus depends on the arbitrariness of a level of description. This point is also made in P. Smith, 'Modest Reductions and the Unity of Science', in D. Charles and K. Lennon (eds.), *Reduction, Explanation, and Realism* (Oxford, 1994), 43.

that in saying x is D and x is C, where 'D' and 'C' are dispositional and categorical predicate terms respectively, we are not saying that there is some fact about x over and above instantiating C that makes it true that it is instantiating D. There is no need, for instance, to say that x has some active power or potency of some completely different kind of property to having C. Such a position would lend credence to Goodman's charge of dispositions being ethereal. The monist wants to say that there is just one attribute of x, or state that x is in, that makes it true of x that Dx and that Cx.

This requirement can be satisfied even if the extensions of D and C do not coincide. Thus there need not be an identity of universals for monism. Being D need not be the same, in every case, as being C as long as each instance of the disposition is identical to some instance of a categorical base. I am relying upon a notion of property instances or states such that instances of particular dispositions and instances of categorical properties can be identified in what amounts to a token–token identity theory.

This sort of move is a common response to variable realization in the philosophy of mind but Prior has argued that this move is not possible for the case of dispositions because the tokens identified in legitimate token–token identity statements are particulars whereas dispositions and categorical properties are not particulars: they are universals.[23] Prior is aware that her opponent may be undeterred at this stage and that they might insist that the identity statements they would offer are not identities between properties as universals, but between property instances that are particulars. Prior handles this line in an unsatisfactory way. She says:

> [I]t may be suggested that what is being advocated is not that this instance of fragility = having bonding α but rather that *this instance* of fragility = *this instance* of having structure β. To allow this would be to allow that instances may be identical when the properties of which they are instances are not identical.
>
> Perhaps, if only because of the obscurity of the relevant notion of 'property instance', this cannot be definitely ruled out. But in any case the thesis that the properties themselves cannot be identified remains untouched, for that thesis is a thesis about properties not instances.

[23] *Dispositions*, 76.

It seems, therefore, that Prior has just two qualms about this type of token–token identity view:

(i) it fails to produce identities of properties; it yields only identities of instances of properties,
(ii) the notion of a property instance is obscure.

Neither of these worries should deter us, nor should the type of property monism we are left with be thought of as weakened or inadequate. Properties *qua* universals are indeed not identified, as Prior says, but they need not be for property monism to be true and dualism false, which is what is at issue. As long as there is some sustainable notion of a property token, then if every disposition property token is numerically identical to some categorical token, there is just one token of a particular with two ways of characterizing it. Thus we can have property monism even if identifications between universals cannot be made. The only problem becomes, therefore, whether there is some plausible notion of a property instance that can warrant our consent.

One candidate notion is that of a trope, being an 'abstract particular' such as the redness of a particular apple or the squareness of a particular window. According to the ontology of tropes, squareness and redness in general do not exist but only a sum of particular rednesses and squarenesses of things.[24] The problem with such an account, however, is that identified by Russell[25] that the trope theorist cannot explain how a number of tropes resemble each other. How is it that ten red things are all red, for instance? To say that they have something in common is to say that they all instantiate the same universal.

Tropes are not necessary for an adequate notion of a property instance, though. For the purposes of property monism for dispositions and their bases we can allow universals in the abstract and realist sense, but such universals would be causally inert. When we say that the weight of the apple caused the pointer on the scales to move, for example, we do not mean that a property of weight in general, construed as a universal, caused the moving of

[24] See K. Campbell, *Abstract Particulars* (Oxford, 1990), and J. Bacon, *Universals and Property Instances* (Oxford, 1996).

[25] B. Russell, 'On the Relation of Universals and Particulars', *Proceedings of the Aristotelian Society*, 12 (1911–12); see also C. Daly, 'Tropes', *Proceedings of the Aristotelian Society*, 94 (1994).

the pointer.[26] Rather it was this particular weight of this particular apple that caused the pointer to move. Similarly, my hair does not possess the colour brown in general; for a universal as traditionally construed does not even have a location, rather it possesses one particular instantiation of that property and it is this instantiation of the property which causes my hair to look brown. Unless we accept some notion of properties being instantiated in particulars, then it seems difficult to sustain the evident link between a thing's properties and the causal transactions into which it enters. It is difficult, in short, to see how, unless we allow that there are particular property instances possessed wholly by objects and substances, a thing's properties can have causal effects. What causes a square peg to fit a square hole? It is not a timeless universal that exists nowhere, rather it is something about this hole and this peg, regardless of what else exhibits a similar quality elsewhere.

This falls far short of justification for the theory of properties I would recommend. That would have to be given elsewhere. What it shows is that some of the traditional philosophical ways of thinking about properties have inadequacies that a move to a notion of a property instance or token would avoid. Prior gives insufficient credit to these notions and refrains from attempting to demystify the obscurity she finds. Given that this was the only remaining objection Prior had to token–token identities between dispositional and categorical property instances, then we had better allow that she has failed to defeat the notion. This means that the argument from variable realization is disarmed because we can now say that the case of dispositional–categorical identities is sufficiently similar to that of psychophysical identities such that the same move, to token–token identities, is available for dispositions in response to the variable realization argument.

This allows further comment on the argument from rigid designation in that the necessity of identity that is required for the identity of two universals is not required in the case of particulars, which property instances are; indeed it becomes an irrelevant question, when two states stand in a relation of token–token identity only. Identities between properties, considered as real

[26] Though Armstrong does defend this line in his *What is a Law of Nature?* (Cambridge, 1983) where laws are real connections between Universals.

universals, are no longer necessary. The putative identities are between individual states of objects or instances of properties.

7.7 *Identity Conditions*

Identity conditions for dispositions and categorical bases are derived from the argument from identity of causal role. Where *d* is a variable to be filled by any disposition token, *c* is a variable to be filled by any categorical base token and *x* and *y* are variables ranging over actual and possible events, the identity conditions for *d* and *c* are:

$\forall d\ \forall c\ ((d = c) \leftrightarrow \exists x\ (d$ causes or is caused by x & c causes or is caused by $x)$
& $\neg\exists y\ ((d$ causes or is caused by y & $\neg(c$ causes or is caused by $y))$
$\vee\ (\neg(d$ causes or is caused by $y)$ & c causes or is caused by $y)))$

The possible events relevant in this formulation are physically possible events; that is, events in possible worlds with the same laws as the actual world. Identity of *actual* causes and effects would not suffice because there is the possibility of *x* and *y* matching in their actual causes and effects though differing in an unexercised causal power and thus in some possible causes and effects.

This itself is a dispositional criterion of property instance identity insofar as is stipulates that any two tokens with all the same causal roles are identical, and therefore are just the one token. There is nothing trivial in giving dispositional identity criteria for the dispositional and non-dispositional, however, for an identity statement derived from these identity conditions is still informative. Although the disposition has its causal powers as a matter of conceptual necessity, a property token classified non-dispositionally does not: its causal powers are contingent upon the contingencies of the laws of nature.

It can be noted also, that the thesis of dispositional–categorical identity has the consequence that a dispositional criterion is not the only criterion of identity. Given property monism, then a criterion of identity for any two property tokens can be given in terms of categorical tokens as follows. For any two dispositional and categorical property tokens, *d* and *c* respectively, then $d = c$ if and only if there is no categorical specification S that belongs to *d* and not c, or to c and not d. Thus identity conditions can also be stated in categorical terms thus:

$$\forall d \, \forall c \, ((d = c) \leftrightarrow \neg \exists S \, ((Sd \,\&\, \neg Sc) \vee (\neg Sd \,\&\, Sc)))$$

which, again, remains informative even though one of the property tokens, c, in the identity statements may have such a categorical specification as a matter of definition. It is important to note that S, in this identity criterion, is not a second-order property possessed by d and c but rather a possible true categorical description of d and c. Again this will yield informative identity statements because although the description S will be true of c trivially, it will true of d only contingently as which dispositions are to be identified with which categorical property instances will be, again, contingent upon contingent laws of nature.

It may be objected that any such identities between dispositions and their bases lead to absurdity because of the possibility of a single categorical base grounding two separate dispositions. If this is a possible case, meeting the identity criteria above, then we could have a situation where $c = d_1$ and $c = d_2$. By transitivity of identity, this means that $d_1 = d_2$, contrary to the assumption that d_1 and d_2 are two separate dispositions. The answer is that while this would be an absurdity, were we talking about universals, in the case of property instances there is no such absurdity in these identity statements as these would be contingent a posteriori identities between particulars. A single categorical base c could well ground two dispositions: one a disposition to cause G_1, if F_1, another the disposition to cause G_2, if F_2. It is not absurd to say that an instantiated property will cause one type of event in one situation and another type of event in another situation. Hence to say that $d_1 = d_2$ amounts to saying that there is a single property instance capable of causing both the kinds of behaviour conceptually connected with those two types of disposition. Looking at it a different way, there can be said to be two tests for the presence of the same property, as when we have various tests for whether a wire is live. Note, however, that on the conditional analysis this could not be allowed because only where we have identical conditionals can we be said to have identical ascribed dispositions. Where we say that a disposition is a property capable of causing certain effects, we *are* at liberty to say that the same property can cause different effects in different circumstances.

7.8 *Supplementary Claims of Property Monism*

The argument from identity of causal role fails to characterize property monism in full. The argument supports the conclusion that any disposition token and its categorical base are to be identified if they occupy identical causal roles, but might property monism be falsified by the existence of a disposition for which there is no categorical base with which it could be identified or by an instantiated categorical property with no causal powers with which it could be identified? This raises the possibility that while there may be some states of objects with both dispositional and categorical terms true of them, there may also be some states that are purely categorical and others that are purely dispositional. This would yield a mixed ontology that the property monist would want to reject. The property monist has, therefore, two supplementary commitments in addition to the argument for identity. These are:

(i) *Every property has some causal role.*

All properties are causally potent, as discussed in Sect. 6.2, in accordance with the causal criterion of property existence. This thesis must be taken as applying to instantiations of concrete properties only. The case of abstract universals will be considered in the next section.

The addition of this commitment to the argument for identity is an assertion that 'purely categorical', causally inert property instantiations are a fiction. Thus, for any property token that can be referred to in categorical terms, there is another description in dispositional terms possible that refers to the same property token. This follows because every instantiated categorical property token is also a causal power.

(ii) *Every causal role has a non-dispositionally specifiable occupant.*

The second thesis concerns a categorical description for any property token that it is possible to denote with a dispositional term. This categorical description is one that is to be discovered empirically and may involve different types of categorical property specification for a single type of disposition, consistent with the variable realization hypothesis.

There is, however, an apparent counterexample to this thesis which requires comment. It is often urged that there are dispositions for which there is no non-dispositionally specifiable occupant of that causal role. Should we take these claims about dispositions without bases seriously? Does it mean that property monism cannot be true?

Two special cases of problematic dispositions have to be considered: abstract dispositions and ungrounded dispositions. I will look at each and consider whether they have a legitimate claim to classification as dispositions and, if so, how property monism can accommodate them into its ontology.

7.9 *Abstract Dispositions*

The example I gave of an abstract disposition was being divisible by 2; an abstract property possessed by even numbers. There are some features this property has which typifies it as dispositional and yet others which distinguish it from dispositions.

The case for divisibility by 2 being regarded as a dispositional property is that, first, being divisible by 2 is something that can be true of more than one number in the same way that being soluble can be true of more than one substance. Second, being divided by 2 is something that can be done to certain numbers in a similar way that being dissolved in water is something that can be done to certain substances. Third, to predicate divisibility by 2 is to ascribe a functional role that a number can play; it is to say that if divided by 2, then a whole number will remain. Fourth, focusing just on the example of divisibility, though it applies also to other mathematical properties, numbers can have different functional roles in different conditions. These may be regarded as corresponding to the different stimlus conditions for a non-abstract disposition where the functional role is a causal role. Hence, a single number may be divisible by two or by three or by five, as in the case of the number thirty.

However, dissimilarities between the abstract case and more paradigmatic dispositions have been pointed out to me.[27] It may be wondered how an abstract disposition fits with the typifying

[27] By Robert Cummins in correspondence.

principle that dispositional properties need not always be manifesting. This is true for paradigm 'concrete' dispositions such as fragility and solubility and yet there seems to be no way in which being divisible by 2 could be sometimes true of a number and sometimes false. The inconceivability of this situation is partly a result of there being no clear idea of what is to count as the manifestation of an abstract disposition. Does a number lie in wait until a divisor comes along? If a divisor does come along, what does it then do? What change occurs that could possibly qualify as a manifestation of the disposition?

These questions spring from the assumption of too narrow an understanding of dispositions: an understanding that restricts them to causal-role occupying properties only. A functionalist understanding of dispositions is one that captures the 'concrete' cases where the functional role is a causal role but more besides. If there is sufficient reason to regard the abstract properties in question as dispositional, and if an account of dispositions, such as the functionalist account, can capture these properties and explain why they are dispositional, at no extra cost to the account, then the more general account is preferable. The Cummins line of attack must be reconsidered, therefore, without an assumption that it must be a causal role that is occupied by abstract dispositions.

I would suggest that being *divisible* by 2 means that a number can be *divided* by 2 to produce a whole number and that such an act of division is to count as the manifestation of the disposition. This kind of manifestation is not to suggest any psychologism about mathematical objects and their properties, though. Abstract dispositions can be accommodated even if numbers are understood as real objects. Being divisible by 2 is a property that certain numbers have constantly and eternally and whether or not there are any sentient beings to perform acts of division upon them. This is perhaps the source of the mistaken thought that divisibility is something that is manifesting all the time. Being divided by 2 is something that is done to a number, in a calculation, and in virtue of that number's eternal mathematical properties; hence being divided by 2 is not always true or false of a number though being *divisible* is. This does not mean that the number becomes involved in a causal interaction with the mind of the reckoner, in which case it might be classified as a 'concrete' disposition in virtue of possessing a causal role. The functional role of the number is the function it can play in our calculation but the connection between the properties of numbers and correct calculations is a rational, logical, and necessary one.

It seems that this kind of functional role provides a plausible account of a non-causal manifestation for an abstract object. Given no other foreseeable problem for accepting this as a legitimate sense of being a disposition, then I suggest that we accept that there are abstract dispositions. It should be added before moving on, however, that the principle cited by Cummins, that dispositions may not be always manifesting, is not itself incontestable. Radioactivity is always manifesting; indeed it is impossible for it not to. Cummins would claim, however, that, for this very reason, radioactivity is not a disposition.

The issue is raised, therefore, of how abstract dispositions fit into the property monist's ontology. The argument from causal role clearly has no application to these cases, as causal roles are not involved. However, the property monist is not committed to monism with respect of all kinds of property. A fundamental split in ontology between 'abstract' and 'concrete' properties could be accepted, for instance. What was at issue in the case of concrete properties was whether there needed to be an extra 'power' added to a particular, over and above its categorical properties, to account for its possible behaviour in various circumstances. We saw how allowing there to be such powers would be a gratuitous double-counting because a single token can instantiate two different properties. An analogous monistic type of response can be given for the case of abstract dispositions. Does there have to be some 'abstract power' added to the abstract properties of a number that make it divisible by 2? Apparently not. To ascribe divisibility by 2 is just to give a functional characterization of that number's properties: it is to say what can be done with it, rather than saying that the number has some *additional* property.

7.10 *Ungrounded Dispositions*

The argument from identity of causal role states that dispositions and their so-called categorical bases occupy identical causal roles and are therefore identical. But what of dispositions which allegedly have no categorical base? These 'ungrounded' dispositions are commonly understood to be the fundamental powers of subatomic particles which, according to prevalent interpretations in theoretical physics, have a causal role but nothing which causes them to behave the way they do. If such dispositions exist, then they are quite different from the other kind of dispositions I have been considering. There is no explanation of why they possess their dispositions; no underlying categorical explanation. They are more like powers in the old-fashioned sense: real potentialities that initiate changes but seem to have a mysterious existence in between these changes.

What are these alleged ungrounded dispositions? Paul Davies suggests 'forces' such as charge, spin, and the radioactive decay of subatomic particles.[28] Different types of particles have different half-lives, charges, and spin and such 'forces' constitute the vital statistics of the subatomic particles. By this I mean that such specifications of charge, spin, and half-life are all that we can say about these things; they are their only properties and, arguably, they are all dispositional properties. Why these particles have these dispositions is inexplicable: there is no categorical base or underlying mechanism that explains their behaviour, for they are supposed to be without structure. Arguably, such ungrounded dispositions will always have to be posited at the bottom of everything because unless something inexplicable is granted, then we will have an infinite regress of explanatory mechanisms.[29]

Are these things really dispositions? Previously I have said that it was questionable whether such behaviour could be attributed to a disposition but the reason I gave was not decisive.[30] Instead, I now think that it makes better sense to allow 'ultimate dispositions' into our considerations and to find a way to accommodate them into property monism. What the property monist should say about such cases, if they exist, is that they are cases where something is missing, rather than cases where something extra is possessed. There is a disposition for which there is no categorical base but we still, arguably, have the disposition. This leaves us with a disposition without a categorical base, which admittedly is strange, but such a case cannot be ruled out a priori. The property monist is

[28] Davies, *Superforce*, 82.

[29] This is an issue I take up in Ch. 10 when I consider two world views.

[30] 'Dispositions, Bases, Overdetermination and Identities', 58.

primarily concerned to argue against the double-counting of a thing's properties. Their strategy is a reductive one, though not a reductionist one,[31] where what were thought to be two properties or states of an object are discovered to be just one state described in two different ways. With an ungrounded disposition we have a property or state of a subatomic particle with just the one mode of characterizing it available to us: the dispositional. We cannot rule out a priori, however, a situation in which a categorical characterization becomes available to us as theory advances. Nowhere is there the sort of property dualism that the property monist is trying to remedy.

This completes the look at property monism though it does not fully determine an ontology for dispositions. Within the confines of a property monism there are a number of ways to turn. Different property monists argue for different positions. What I will do in the next chapter is consider these different positions and advance my own form of monism.

[31] See ch. 8 for this issue.

8

Eliminativism and Reductionism

8.1 *Reductionist and Eliminativist Monisms*

There are various positions which could be called property monist as so far defined. There is still work to be done, therefore, before a complete ontology for dispositions has been set out. This is apparent from the number of different positions that have been defended by the protagonists in the dispute that could be interpreted as property monist.

My aim in this chapter is to narrow down the kind of property monism we ought to support. I will do this by presenting the alternatives, weighing up the advantages and disadvantages of each, and then pointing towards my own version of monism which differs from all of these.

First, though, we need to consider how many possible choices are on offer. One distinction that is relevant to this question is one that has largely been ignored in the debate. This is the distinction between property monisms that are reductionist and property monisms that are eliminativist. It is notable that ontologies that give some kind of priority to the categorical and those that give it to the dispositional are not always diametrically opposed. While one may be an eliminativist position, its opponent may be reductionist and thus it will be making a claim of an altogether different nature. The protagonists have not in general been sensitive to this distinction, even though it is substantial, and so certain key points of contention have been oversimplified or ignored. The previously proposed twofold division of monistic ontologies into dispositional and categorical monisms is thus not enough.[1] To be sensitive to important claims we need to divide each of these types of

[1] This distinction was the one used in my 'Dispositions, Bases, Overdetermination and Identities'.

monisms in two, giving us reductionist and eliminativist varieties of each and thus yielding an initial four possible positions.

To justify this fourfold division we need to consider what the differences are between reductionist and eliminativist positions. We can start to understand this by examining Quine's pejorative comments on dispositions and considering his response. A typical comment comes when he says:

> Advances in chemistry eventually redeem the solubility idea, but only in terms of a full-blown theory. We come to understand just what there is about the submicroscopic form and composition of a solid that enables water to dissolve it. Thenceforward, solubility can simply be equated to these explanatory traits. When we say of a lump that it would necessarily dissolve if in water, we can be understood as attributing to the lump those supposedly enumerated details of submicroscopic structure—those explanatory traits with which we are imagining solubility to have been newly equated. A chemist can tell you what they are. I cannot.[2]

Elsewhere he adds to this:

> Each disposition, in my view, is a physical state or mechanism. A name for a specific disposition, e.g. solubility in water, deserves its place in the vocabulary of a scientific theory as a name of a particular state or mechanism.[3]

Five important claims are being made in these passages:

1. Dispositions are ideas or concepts that stand in need of redemption.
2. An understanding of a phenomenon in dispositional terms is not an adequate understanding of that phenomenon for scientific purposes.
3. Non-dispositional terms should replace the dispositional idiom.
4. It is a job for science to replace the dispositional idiom. The physical details that scientists discover can 'even take the place of the old disposition term, or stand as its new definition'.[4]
5. (By implication) if there is no particular state or mechanism that a disposition term can be equated with, then the disposition term does not deserve a place in scientific vocabulary.

There are obviously a number of questions raised by Quine's claims. Why do disposition terms stand in need of redemption? Why, if

[2] *Ways of Paradox*, 71–2. [3] *Roots of Reference*, 11. [4] Ibid.

dispositions can be equated with these physical mechanisms, by which one assumes Quine means that they can be identified with them, are the mechanisms more acceptable than the dispositions? Is this merely a part of Quine's attack on intensional idioms in general or is it also a result of his philosophical naturalism? Is it the case that Quine thinks that there really are no dispositions, *qua* properties, but merely a dispositional idiom for talking of such properties and that all such properties are in reality non-dispositional?

There are two possible directions the Quinean can take given this pejorative view of dispositions: reductionism or eliminativism. Disposition terms are claimed to have no role in scientific discourse. If they can be equated, one-for-one, with terms from a scientifically acceptable vocabulary, then Quine's stance indicates reductionism about the dispositional. Each disposition term is reduced to a non-dispositional one. However, the talk of disposition terms deserving their place in the scientific vocabulary only *if* they can be identified with a non-dispositional mechanism suggests the possibility of disposition terms for which no such identification is possible. These disposition terms are beyond exoneration and thus are unfit for incorporation into science. The Quinean response to these terms is presumably that they should be eliminated completely. No such property alluded to by the disposition term exists. This kind of case provides us, therefore, with an example of an eliminativist attitude towards dispositions.

My purpose in this chapter is not to uncover the correct interpretation of Quine, reductionist or eliminativist, because I am not sure that it is an issue he is sensitive to and there is no reason to expect a considered and consistent response from him. The approach I will be taking is one where I will explore the different claims of the eliminativist and reductionist responses and provide some assessment as to their relative merits. I begin by attempting to give an explicit formulation of each of the four possible positions. Not until I have considered these will I present my alternative.

8.2 *The Varieties of Property Monism*

The four initial types of property monism I call categorical reductionism, dispositional reductionism, categorical eliminativism, and dispositional eliminativism.

If we are to say that there are not separate dispositional and categorical properties, that is, if we are to reject property dualism, then there are two ways we can do this. The first is to say that one type of property is reducible to the other, with identities between universals. The other is to say that one type of property has no existence at all: that it is eliminable.

Essential to the reductionist position is an identity relation between dispositional and categorical properties or property instances.[5] This is what I have elsewhere called the coextension requirement.[6] If one class *K* of properties is to be reducible to another class *J*, then members of *K* must have identical extensions or referents as the members of *J*. This I take to be a minimal demand on reduction for, while it does supply some kind of reduction, coextension alone fails to determine the reduced and reducing classes. The minimal kind of reduction provided by satisfaction of the coextension requirement is that where previously there were thought to be two properties that do the causal work in a disposition manifestation there is actually only one. Thus, the list of separate causally relevant properties in our explanation of an event is reduced.

Reductionism, in its stronger varieties, however, urges more: some kind of priority of one class over another is being claimed. This reductionism is distinguished by there being a putative direction of reduction: one class is asymmetrically reduced to the other. The reduced class must be somehow subsumed by, or be incorporated into, the reducing class in a way that the reductionist will have to explicate. In my earlier discussion of this question I went on to dismiss one possible way in which reduced and reducing could be determined but I now think there are other more plausible ways in which this can be done and I will discuss these options in what follows.

Categorical and dispositional reductionisms allow that each disposition is identical to some categorical property or categorical property-complex in the sense described in the last chapter. The two ontologies can be separated in the following way:

[5] I am allowing either type–type identities of properties or token–token identities of property instances. The arguments presented will be applicable to either case though for ease of expression they will usually be expressed only in terms of one of the alternatives.

[6] 'Dispositions, Supervenience and Reduction', 421.

(*a*) *Categorical reductionism* is the position that all dispositional properties are reducible to categorical properties.

My purpose is not to discern exactly where each philosopher fits in with respect to this classification but I will offer Armstrong as representative of categorical reductionism when he says 'to speak of an object's having a dispositional property entails that the object is in some non-dispositional state or that it has some property (there exists a "categorical basis") which is responsible for the object manifesting certain behaviour.'[7] It becomes clear, as Armstrong describes his account further, that the dispositional property is to be identified with its categorical basis rather than the relation being solely that of entailment. I think there is a clearer case for classifying Armstrong as a categorical reductionist than there is for so classifying Quine because Armstrong states explicitly that every dispositional property entails a categorical basis to which it is identical. I think it is also clear that Armstrong thinks that there is an asymmetrical relation holding between dispositional and categorical pairs such that it is the dispositional that is reduced to the categorical and not vice versa. He speaks, for instance, of scientists discovering the 'concrete nature' of the disposition.[8] He seems also to suggest that everything real is categorical and there can be no admission of real potentialities. He says that 'it is impossible that the world should contain anything over and above what is actual. For there is no mean between existence and non-existence.'[9] This suggests that there is some direction of reduction that Armstrong wants to support: some sense in which all properties in the world are categorical and in which every dispositional property can be reduced to a non-dispositional one.

(*b*) *Dispositional reductionism* is the position that all categorical properties are reducible to dispositional properties.

This position is one that accepts an identity relation between the dispositional and the categorical but differs from categorical reductionism in respect of the direction of reduction. It is, however, difficult to find a representative of this position. While there seem to be cases where the categorical is regarded as respectable

[7] *A Materialist Theory of the Mind*, 86.
[8] 'C. B. Martin, Counterfactuals, Causality and Conditionals', 14.
[9] 'Dispositions are Causes', 24.

enough to be taken as real and the dispositional real only if identical to a categorical property, few have found it tempting to argue that the categorical is real only insofar as it is identical to a dispositional property. There are plausible reasons why no one has been attracted by this position. To argue for dispositional reductionism would be to regard the dispositional as a legitimate ontological category. Those who do so, however, may find other views more compelling as a result. Most dispositional monists are actually dispositional eliminativists, for example, as we shall see. Another response to the acceptance of dispositions as legitimate would be property dualism. Thus there is a significant inequality in the popularity of the two reductionist positions. In due course I will argue that neither option ought to be regarded as persuasive. This leads on to the eliminativist positions:

(*c*) *Categorical eliminativism* is the view that all properties are non-dispositional and there are no properties corresponding to disposition terms.

I think it is a good idea to separate this claim from that of the conditional analysis, though there are obvious claims in common. The conditional analysis involves the denial of the existence of dispositional properties, in common with eliminativism, but it is connected with an event ontology where properties in general are viewed with suspicion. Properties are instantiated only in events and, for the committed empiricist, have no endurance between such events. I have already discussed and dismissed this kind of account for dispositions. The kind of eliminativism I will address here, however, is that where the ontological category of *property* is accepted: there are real instantiations of properties enduring between events. The position states, however, that these are all categorical and there is no dispositional property in reality that a disposition ascription denotes. We have seen, with the example of Quine's views, the sorts of reason that could be given for this view. Quine thought that if there is no categorical basis that supports a putative disposition, then that disposition term can be, and should be, eliminated from our vocabulary.

(*d*) *Dispositional eliminativism* is the view that all properties are dispositional, even those which are traditionally regarded as paradigmatically categorical, and there is no such thing as a categorical property.

The most obvious proponents of this view are Popper and Mellor.[10] Their argument, as discussed in Chapter 4, is that all property ascriptions are conditional-entailing, and given that conditional entailment is the most obvious mark of the dispositional, then all properties come out as dispositional. The claim here is that the notion of a categorical property is a fundamentally misconceived one. The very idea of a categorical property contains the impossibility of a non-conditional-entailing property ascription and so, as a matter of necessity, no property can meet the conditions for being categorical. All properties, even those traditionally understood as categorical, are thus, in truth, dispositional.

The following generalizations about the four positions can be made. Categorical and dispositional reductionists agree over which properties exist but disagree over the direction of reduction between the two classes of property terms. Categorical and dispositional eliminativists agree that we are mistaken in thinking a certain class of properties to be real but disagree over which class of properties they are. This dispute does not essentially involve the existential status of particular properties. For instance, dispositional eliminativists need not deny that there is such a property as triangularity. Rather, the dispute may be a conceptual one of how we are to classify such a property: dispositional or categorical. According to Mellor, for instance, our understanding of the notion of a dispositional property is broad enough such that it encompasses both solubility and triangularity and any other truly predicable property. In some varieties, however, the eliminativist may deny the existence of a particular property. The categorical eliminativist may deny, for example, that fragility is a real property at all. They may claim that a God's-eye view of the world would detect nothing that corresponds to this term. Obviously this is a controversial claim that demands further attention. Having made the initial distinctions, therefore, I now move on to consider in detail the reductionist and eliminativist cases, beginning with eliminativism.

[10] Popper, 'The Propensity Interpretation of the Calculus of Probability', and *Logic of Scientific Discovery*, appendix X, and Mellor, 'In Defense of Dispositions', and 'Counting Corners Correctly'.

8.3 *For Eliminativism*

What essentially distinguishes an eliminativist position from a reductionist one is a denial of identities between dispositional and categorical properties. This is a position that was originally developed in the philosophy of mind.[11] The reason we cannot say that mental events and physical events are identical is, according to the eliminativist, that the language of our psychological vocabulary is such a fundamentally misconceived one that it bears little, if any, resemblance to what really exists. Hence we cannot say for every ascribed mental event that there is some physical event to which it has a one-to-one correspondence and is thus a candidate identical event.

Examples can be given from earlier explanatory vocabularies that are more obviously redundant. Explanations of mental illness in terms of demonic possession have not been reduced to a language of psychological and neurophysiological disturbances: they have been eliminated. Demonic possession has not been accommodated in our current theories because it is thought to be a fundamentally misconceived explanation that appeals to causal agents that have no existence whatsoever.

An eliminativist position in the case of the dispositional or the categorical will similarly involve the claim that one type of property has no existence and so is identical with no existent property. The position thus remains property monist but makes the claim that one category of properties is to be dispensed with rather than incorporated. I will describe first the categorical eliminativist's argument and then the dispositional eliminativist's.

A. Categorical eliminativism

One view of dispositions that may suggest a categorical eliminativist conclusion is the view that they are ethereal or occult powers that have a pre-scientific status and cannot be expected to match up, one for one, with the actual mechanisms that are

[11] P. Feyerabend, 'Mental Events and the Brain', *Journal of Philosophy*, 60 (1963), gives an early statement of eliminativism in the philosophy of mind. See P. Churchland, 'Eliminative Materialism and the Propositional Attitudes', *Journal of Philosophy*, 78 (1981), for a fuller presentation.

found scientifically to be their *explananda*. Just like demonic possession, the language of dispositions relies on entities which simply do not exist. The categorical eliminativist claims that there are no dispositional properties. All properties must be conceived of as categorical states or mechanisms in the spirit of Boyle's explanations of powers.

There are a variety of reasons why the categorical eliminativist regards dispositions as objects of suspicion. They may be thought, as in Goodman, to be appeals to ethereal or ghostly forces that typify the explanations of superseded pre-science. If a disposition term deserves its place in scientific theory only as a promissory note to be redeemed by some physical state or mechanism, presumably there are disposition terms that do not correspond to any physical state or mechanism and so do not deserve any place in a scientific vocabulary. Any tight correspondence between a disposition term and a type of physical mechanism must surely be thought of as nothing but a stroke of good fortune on the part of our, largely ignorant, predecessors. The dispositional idiom was around for a long time before we began to understand the true causes of changes in things and we would have had to be very lucky to have created a pre-scientific vocabulary whose terms matched up accurately with those scientists would use now. We cannot expect that our early explanations of change in the world, formed when there was little understanding of the actual processes that occurred, were so close to the mark that re-description of exactly these properties suffices. The eliminativist will claim that the odds of having provided an accurate description of reality at our first attempt are astronomical.

Disposition terms fail to denote properties because there is no single type of state that is possessed by things to which we ascribe them. We may ascribe 'solubility' to various substances but this term does not uniquely identify any real property of such a substance. Those things which we in practice call 'soluble' will have a variety of different molecular structures that are responsible for a variety of different behaviours. Note, as Prior says, not everything that is soluble is soluble to the same degree or soluble in the same circumstances.[12] Put another way, we could say that everything will be soluble if we find some peculiar set of conditions with

[12] *Dispositions*, 1.

which to subject that thing. What this suggests is that calling things 'soluble' is a lazy and inaccurate way of talking about their properties that no self-respecting physical scientist would ever use.

Quine mentions the disposition of intelligence as an unredeemed promissory note. Those who have an eliminativist response to Quine may be sceptical that 'intelligence' can ever be redeemed. Not only is it that all sorts of brain processes are likely to be implicated in intelligence but also the disposition term itself is extremely vague. We are not sure what abilities we are referring to when we describe someone as intelligent. At best it seems to be some large disjunction of abilities that qualify one as intelligent for it is controversial that a single test, such as an Intelligence Quotient test, can be an accurate measure of intelligence. The eliminativist may, therefore, happily allow the disposition term to be eliminated in favour of more accurate descriptions of types of neural network, for instance.

B. Dispositional eliminativism

What are the claims that typify the dispositional eliminativist position? There must be at least two claims. First, that specifically dispositional properties are real, that is, dispositions terms refer to actual properties rather than being terms elliptical for a (set of) conditional(s). A conditional analysis reduces dispositions away into complexes of events but the dispositional eliminativist is making the claim that all properties are dispositional—there exist no non-dispositional properties—so they must make the claim that dispositions are something more substantive than the conditional analysis permits. Second, the eliminativist must claim that all apparently categorical properties are actually dispositional properties. This is done by supporting the view that all, including even 'paradigm', categorical properties meet the criterion of dispositionality, that is, the criterion by which a property is classed as a dispositional property.

An asymmetry between the categorical and dispositional eliminativist positions can be noted here. Whereas a categorical eliminativist may deny that there is any such thing as elasticity, intelligence, or dormitive virtues, the dispositional eliminativist can hardly claim that there is no such property as triangularity or squareness. The dispositional eliminativist instead is typically someone who argues for a reclassification of these properties as dispositional. In other words, they argue that no property meets the conditions for being categorical.

Mellor, in accordance with a common view, appears to take conditional entailment to be the criterion for dispositionality.[13] Mellor is thus in agreement with Popper where he says that 'all universals are dispositional'. Popper then adds that 'universals can be dispositional in varying degrees'.[14] It is not made explicit how it is possible to have different degrees of dispositionality but examples are offered, such as 'soluble being clearly dispositional to a higher degree than "dissolved"'. 'Dissolved' is nevertheless dispositional because 'A chemist would not say that sugar or salt has *dissolved* in water if he did not expect that he could get the sugar or salt back by evaporating the water.'[15] 'Dissolved' thus entails the conditional 'if the liquid is evaporated, the solid is recovered'.

However, conditional entailment as the mark of the dispositional is only a contingent feature of dispositional eliminativism. Something else may be taken as the mark of the dispositional,[16] but for categorical properties to be eliminated they must be shown to be dispositional in virtue of satisfying the criterion of the dispositional.

In summary of the dispositional eliminativist case, it is argued that all real properties meet the criterion of dispositionality. Typically, though not essentially, this is taken to be conditional entailment, thus there are no properties that are categorical.

8.4 *Against Eliminativism*

The question to be answered now is how strong are the eliminativist positions. I will argue that neither form of eliminativism is strong enough to warrant support. My reasons follow from the analysis of dispositions I developed earlier. A problem with each version of eliminativism is that it starts with a misconceived view of what a disposition is. I will first make this clear for categorical eliminativism.

[13] 'Counting Corners Correctly', 96.
[14] *Logic of Scientific Discovery*, 424.
[15] Ibid.
[16] U. T. Place, 'Intentionality as the Mark of the Dispositional', *Dialectica*, 50 (1996), suggests a different approach but one which cannot be taken as convincing.

The mistake in the categorical eliminativist's claim that a disposition term fails to designate a single property appears to be based on the taking of categorical criteria as criteria for the dispositional.

I conceded that dispositions could be variably realized by different categorical properties (Sect. 7.6). This means that a number of different categorical property predicates can be true of a particular for which a single disposition predicate is true. However, to move from this true claim to the false claim that there is no single property true of all particulars which are ascribed a disposition predicate D is invalid. The error is to assume that it must be a single *categorical* predicate that is true of all such particulars for it to be true that there is a single property in common. This involves the tacit assumption that the single property in common cannot be a dispositional property. This would be a claim justified only once eliminativism had been accepted and so it cannot be used in a proof of eliminativism.

A disposition term 'D' designates a functional property. If my analysis has been correct, then the single property that all particulars that satisfy 'D' have in common is a functionally characterized property that can cause *G*, if *F*, in ideal conditions *C*. By freeing ourselves from the assumption that it must be a categorical property that is in common we can see that it is very difficult for any eliminativist to maintain either that such a functional specification does not exist or that such a property is not common to all particulars which satisfy 'D'.

The example of the property of being red, construed as a dispositional property, illustrates the issue. Let us assume that being red is a disposition that can be variably realized by different categorical bases. Assuming further that a disjunction of properties cannot itself count as a property, should we infer that redness does not exist because its ascription fails to pick out a single property? I argue not. An ascription of redness, on these assumptions, fails to pick out a single categorical property but that is not what is being claimed by a disposition ascription. The criterion for being a disposition of a certain kind is such that it is satisfied by anything that plays the appropriate functional role. Hence there is something in common to all red things: having such-and-such a causal role. Acceptance of this point means that it would not be at all remarkable that the disposition terms used by our predecessors were still of use now. Not only could we say that they had use of a vocabulary of categorical terms which is still of use now, such as 'being square' or 'being water', but also that a vocabulary of useful dispositional terms could be created without any particularly sophisticated scientific background. In calling things soluble we are saying only that they are able to do a particular thing—dissolve in liquid—we are not making any claims about a specific chemical structure that is implicated; hence the fact that no single such structure exists would not undermine the legitimacy of the disposition term's ascription.

What of the problem case of 'intelligence'? Is this a disposition term that does justify elimination because it is a term where even a functional specification of the property is unclear? Admittedly 'intelligence' is a vaguely defined disposition term. It is not clear what abilities are to be had to qualify as intelligent. However, I allowed cases where disposition terms could be analysed only in terms of a disjunction of conditionals (Sect. 3.2). This can now be presented in a way that is free from the conditional analysis and which may show that 'intelligence' can be redeemed. To be intelligent is to have a number of different abilities in different strengths. Whether or not one of these abilities is possessed or whether it is possessed in sufficient strength may be an issue that judgement is required to settle. However, these abilities can be thought of as belonging to a family—there is a family resemblance between them—which is thought to have a causal origin in some fortunate set of circumstances in the brain. The intelligence idea is of some such vaguely defined, functionally characterized property-complex or mechanism. Admittedly the conceptual analysis of intelligence is disjunctive and fairly open-ended, but the idea gains its authority from a conviction that there is something there, in the brain of the owner, that is responsible for these various functionally defined abilities. It is this conviction that makes us refrain from elimination of the term.

However, circumstances in which a disposition term deserves elimination are imaginable. To say that something has a disposition to emit phlogiston when alight is, for instance, to explain the evident behaviour with reference to a disposition that does not exist because there is no phlogiston, hence nothing can have the disposition to emit it. This disposition term can rightly be eliminated but, for the reasons I have given, this does not warrant the elimination of the whole dispositional class of terms.

If we are clear about what a dispositional property is supposed to be, then the credibility of categorical eliminativism melts away. Dispositions are eliminated as a whole class only if they are misunderstood. Properly understood, their elimination has no necessity.

I argue in the same vein that dispositional eliminativism, as I spelled it out, is based on a mistaken conceptual distinction between the dispositional and categorical. As I argued in Chapter 3, the dispositional–categorical conceptual distinction can be maintained even if both types of property ascription support conditionals in some way. The point was that they supported conditionals in different ways. Which particular conditional was supported by a categorical ascription was a contingent matter because it is contingent, for any categorical ascription such as a molecular-structural ascription, what events would follow in any particular situation. This follows from the Humean principle of the logical contingency of causal relations. However, the conditionals for disposition ascriptions follow by analytic necessity because it is part of the meaning of a disposition term that it is a property which causes a particular manifestation if certain conditions are realized. If this analysis is correct, as I have argued it is, Mellor and Popper are wrong to move from the support of conditionals by all properties to a denial of a conceptual distinction between the dispositional and the categorical and a collapsing of the categorical into the dispositional. This means further that not all properties are dispositional and therefore there is no justification of dispositional eliminativism along these lines.

I have argued against the eliminativist construals of property monism and so I now move on to the forms of monism that are consistent with the argument for identity presented in the previous chapter: categorical and dispositional reductionism.

8.5 *Reductionism: The Arguments Assessed*

I argued in Chapter 7 that an identity theory of dispositions and their categorical bases is a stronger position than property dualism. An identity relation is a minimum requirement for the reduction of one class of property denoting terms to the other but it does not suffice for the kind of reduction that the dispositional and categorical reductionists want. Instead, some additional relation will have to exist between the two classes; something that adds the necessary asymmetry that will determine an asymmetrical relation between reduced and reducing.[17]

On its own, an identity theory for two classes of properties does give us some form of reduction: reduction in the sense that where we previously may have thought that two properties existed, a disposition and its putative categorical base, the identity theory makes the claim that there is just one. Hence we have a reduction in the number of types of properties required in our ontology. However, the reductions that I will consider here are those that go further than this and attempt to add some claim of priority for one or other of the classes of property. Such a claim would add to the minimally reductive claim a more substantive and, I will argue, controversial one: that it is the dispositional that is reduced to the categorical or, alternatively, the categorical that is reduced to the dispositional. These more substantive positions concerning the reducibility of one class to the other I shall refer to as *reductionisms.* Thus we have the position of categorical reductionism where the dispositional is reducible to the categorical and dispositional reductionism where the categorical is reducible to the dispositional. I will reserve the term *reduction* for the minimally reductive claim of identity alone.

How could the stronger thesis of reductionism be supported in either the dispositionally or categorically reductive cases? To clarify the answer to this question I will make a distinction similar to the one Nagel made in his classic discussion of reduction: of formal and non-formal conditions for reductionism of one theory to another. The distinction I make will be between internal and external conditions.

Nagel described his distinction as being merely between 'matters that are primarily of a formal nature . . . [and] questions of a factual or empirical character.'[18] I want to be a bit more precise than this and distinguish the relations that hold exclusively

[17] This point was the start of my 'Dispositions, Supervenience and Reduction'. What I have to say here is intended as an improvement on the earlier discussion.
[18] E. Nagel, *The Structure of Science* (London, 1961), 345.

between the two terms '*c*' and '*d*' or classes of terms '*C*' and '*D*', involved in the reductionist statements, from the wider considerations involving matters extra to the relations between the reduced and reducing terms or classes. Relations of the first kind I call internal conditions for reductionism; issues of the second kind I call external conditions. The exact nature of this distinction will become clearer as I progress.

First, what can be said of the internal conditions for reduction? What we are looking for here is whether we can determine reducing and reduced among dispositions and their categorical bases by considering just the relations between them.

Obviously the first relation that we can say would have to hold between pairs, for reductionism to be true, is that of identity, or what I have called the coextension requirement for reductionism. Having discussed what this means at length and having supported a version of it in the previous chapter I shall not repeat that here. It does not suffice for reductionism, being a symmetrical relation.

If the identity relation does not suffice to bring reductionism, is there any other relation, internal to a pair '*c*' and '*d*', that does so? Clearly there is not. There is no relation substantial enough to bring the sort of order of priority required for reductionism that is consistent with the claim of identity. If the required priority is supposed to mean that categorical properties exist and dispositions do not, or that any substantive asymmetrical relation holds between dispositions and categorical bases, then these relations will be inconsistent with the relation of identity, by a simple application of Leibniz's law. This argument rules out any other possible internal relation between pairs '*c*' and '*d*', or classes '*C*' and '*D*', such as supervenience, which it may have been hoped could provide the asymmetry that determines the reducing and reduced classes.[19] I move on, therefore, to the possible external considerations that could determine reducing and reduced.

There are four possible arguments for categorical reductionism, of which I am aware, that are based on wider considerations. These are:

1. *The wider scope of the categorical family of predicates.*
2. *The variable realization of the dispositional by the categorical.*

[19] This simple point takes the place of much of the detailed argument advanced in my 'Dispositions, Supervenience and Reduction', Sects. 3 and 4.

3. *Categorical properties being first-order; dispositions second-order.*
4. *Categorical properties being more explanatorily basic.*

Arguments 1 and 4 could reasonably be turned round to become arguments for dispositional reductionism, as I shall explain. I will consider the merits of each argument in turn.

Argument 1: One argument for the reduction of the mental to the physical in the philosophy of mind concerns the relative 'scope' of the two theories. Physics, it is urged, applies to everything in nature but psychology applies only to one part of it. In other words, all events have physical descriptions but not all events have psychological descriptions. The reduction of the mental to the physical is thus justified on the basis that although every mental event also has a physical description, some physical events have no psychological description; they are identical to no psychological event. The claim can be represented as follows, where Mx iff x satisfies a mental description and Px iff x satisfies a physical description:

[R_1] $\forall x\,(Mx \rightarrow Px)\;\&\;\Diamond(\exists y\,(Py\;\&\;\neg My))$.

Can a similar argument concerning the relative 'scope' of the dispositional and categorical vocabularies be constructed? Could we justify the claim:

[R_2] $\forall x\,(Dx \rightarrow Cx)\;\&\;\Diamond(\exists y\,(Cy \rightarrow \neg Dy))$

where everything with a dispositional description also has a categorical description but not vice versa? This could be cited as support of categorical reduction but, as a form of argument, could also be used to support dispositional reductionism if more plausibility is granted to:

[R_3] $\forall x\,(Cx \rightarrow Dx)\;\&\;\Diamond(\exists y\,(Dy \rightarrow \neg Cy))$.

The categorical reductionist may argue that although all dispositions are identical with their categorical bases—for every disposition there is some categorical property or set of categorical properties to which it is identical—the same does not hold contrariwise: it is not the case that for every categorical property or complex of categorical properties there is some disposition to which it is identical. However, the arguments advanced so far suggest that this claim cannot be sustained. Every categorical property affords causal possibilities to the particulars that instantiate it, as entailed by the causal criterion of property existence (in Sect. 6.2) which was accepted as the most plausible criterion for the reality of a concrete property. This means that everything that has a categorical ascription true of it will also have a disposition ascription true of it. This argument for categorical reductionism is therefore ruled out.

The dispositional reductionist could likewise attempt such an argument but claim that in this case there is no reason to think that all dispositions can also be denoted in categorical terms. Such exceptions could be those putatively ungrounded dispositions that I discussed in Sect. 7.10. However, if appeal to these cases is all that the dispositional reductionist's case rests upon, then it is not a wholly conclusive case. First, ungrounded dispositions are an atypical case and I think it would be unwise to have a theory of dispositions dictated by features of an atypical case. It would be a mistake to say, for instance, that other more commonplace dispositions, such as flammability and elasticity were ungrounded. In all these other cases it is accepted that anything that can be denoted in a dispositional manner can also be denoted in a categorical manner so in these typical cases there seems nothing to compel dispositional reductionism. This is an admission that R_3 may be true but dispositional reductionism still not entailed. I suggest that the move from R_3 to such reductionism can be made only where there are enough mainstream dispositions, the more the better, that have no categorical description. Second, R_3 cannot be known true a priori even if we appeal to some ungrounded dispositions in practice. For any dispositions currently claimed to be ungrounded, we cannot rule out the possibility that a categorical ascription for it will be found as theory progresses. The acceptance of R_3 would, therefore, be a contentious matter.

Argument 2: The mere fact of variable realization of the dispositional by the categorical may be thought sufficient to warrant the reduction of the dispositional to the categorical. The basis for this judgement must presumably be that variable realization shows that the categorical has priority in virtue of holding the asymmetrical relation of *realizing* to the dispositional. The asymmetry of this relation means that while a disposition of type D_1 could be identical in its instances to categorical instances of differing types, C_1, C_2,

C_3, . . . , the converse relation does not hold. The variable realization of D_1 by different categorical bases thus exhibits the following asymmetry:

$$\forall x\ (C_1 x \rightarrow D_1 x)\ \&\ \neg(D_1 x \rightarrow C_1 x).$$

Does variable realization warrant reductionism? Properly understood, it does not. Variable realization is not, for instance, the assertion of some causal determination of the dispositional by the categorical which may be thought to suggest the priority of the categorical. What we have is a relation of identity between dispositional and categorical tokens which, therefore, cannot be related causally. Still, it may be thought, there must be some sense in which variable realization supports a claim that it is *because* $C_1 x$ that $D_1 x$ and not vice versa. This too would be a misunderstanding of the variable realization claim. Although it is true that $(C_1 x \rightarrow D_1 x)$, this cannot be interpreted as supporting the claim that $D_1 x$ *because* $C_1 x$. $C_1 x$ is instantiated in token *i* of $D_1 x$, $C_2 x$ is instantiated in token *ii* of $D_1 x$, and so on. These cannot be understood as relations of determination between the categorical and the dispositional but, rather, facts of identities. That which is (causally) determined by the categorical base of the disposition is the manifestation of the disposition, not the disposition itself. It is thus unwarranted to assume that the categorical base is a *reason* for the disposition. That a particular categorical token is identical with a particular disposition token must be interpreted as a brute fact of nature.[20] It must be concluded, therefore, that although some asymmetrical relation can be found between dispositional and categorical types, this asymmetry is not one that authorizes reductionism.

Argument 3: A third argument for categorical reductionism may be developed from Prior's understanding of dispositional properties as second-order properties. Categorical properties, being first-order, are in some sense prior to second-order dispositions. I do not accept this argument for reductionism because I do not accept the suggested relationship between the dispositional and the cate-

[20] One way of understanding the situation may be that it is because of laws of nature that the identities exist. Although I agree that it is some inexplicable basic fact that must be appealed to, I support an alternative to such general laws, in ch. 10.

gorical. I understand both the dispositional and the categorical as first-order and so on an ontological level-footing.

Prior says that to have a disposition is to have a functional property and to have a functional property is to have 'a second-order property as opposed to a first-order one'. It is: '*the property of having a property that plays a particular causal role*'.[21] On Prior's assumption of distinctness, an understanding of dispositions as second-order brings its own particular difficulties[22] but it is worth considering whether it is still right to regard dispositions as second-order in isolation from the question of the truth of property dualism.

I think dispositions being second-order properties is counter-intuitive in the following way. Properties, including dispositional properties, are ascribed to things. Our basic ontological entity is the object, or a *substance*, to which properties are ascribed. With the Prior view, we are asked to accept the existence of second-order properties, which are properties of having properties and this is contrary to our ordinary concept of a disposition.[23] We ascribe dispositions to objects, not to properties: we say that the vase was fragile and the sugar was soluble, not that the molecular bonding was fragile and molecular composition *xyz* was soluble; for we ascribe these disposition terms almost always in complete ignorance of such microstructural properties. To say that something has a functional property is thus not to say that it has a property of having a (first-order) property. It is to say that it has a first-order property but a first-order property for which the causal role can be known a priori from the meaning of the predicate. Categorical properties do not vary in status with respect to ordering of properties, they vary only insofar as their causal role can be known only a posteriori. There is, thus, here no relevant difference on which to base a claim of reductionism. I accept that much of this is at odds with thinking on functionalism but in the next chapter I will describe more fully the type of functionalism that would be required by this view of dispositions.

Argument 4: A final argument that can be used for reductionism comes from a claim that has already been considered and dismissed,

[21] *Dispositions*, 81. [22] See Sect. 5.8.

[23] It is also contrary to Armstrong's position, *A Theory of Universals*, 143–6, that the only properties of properties are formal.

namely, the claim of explanatory asymmetry between the categorical and dispositional (Sect. 5.3). The argument was used as an argument for property dualism. As an argument for that conclusion, explanatory asymmetry was dismissed but could it be used for a 'weaker' conclusion of reductionism? Clearly not, because the credibility of such explanatory asymmetry was cast into doubt. A disposition can be explained in terms of some categorical property but, likewise, a categorical property can be explained in terms of some further dispositional property. Both types of ascription have explanatory roles to play and it seems that both are suited to explanations of properties from the other category. Thus it seems nothing conclusive about the direction for reductionism follows.

8.6 *Monism without Reductionism*

Why is reductionism tempting when the arguments in its favour are inconclusive? I would suggest that reductionists are manifesting prejudices. Quine's contempt for dispositions, for instance, manifested in his statement that they stand in need of redemption, is a typically empiricist statement where disposition terms are to be dispensed with or replaced wherever possible.

But what is left if there is insufficient warrant for any form of reductionism? My answer is that monism is as far as we need go. Reductionism in either direction is both unnecessary and an incorrect way of looking at the relationship between the dispositional and the categorical. What we have are two different ways of denoting the same properties. One way is in terms of what possibilities those properties bestow on their possessors. The other incorporates a wider class of descriptions involving features such as shape, structure, and molecular structure. I can find no reason why the dispositional should be reducible to the categorical, nor vice versa, given that the dispositional and categorical are correctly understood just as two modes of presentation of the same instantiated properties. Admittedly, we may still speak of dispositional properties and categorical properties as if they were names for distinct properties, but correctly understood it can be seen that we would more accurately speak in terms of the dispositional and categorical styles of denoting properties. Given this understanding of the distinction, there seems no necessity nor use in trying to push monism towards reductionism.

The position that allows equal weight and importance to both modes of property denotation is one that could be called *neutral monism*. This term intends to convey the identity between an instantiation of a disposition and its categorical base. It also conveys the view that properties themselves are just properties *simpliciter*, which should be thought of neither as 'really' categorical nor 'really' dispositional, but which can be denoted in those ways. Accepting this point is a major step on the way to the correct understanding of dispositions.

In the next chapter I will detail the relationship between the two modes of denoting properties in more depth.

9

A Functionalist Theory of Dispositions

9.1 *Preliminaries: Non-Linguistic Reality and Descriptive Idioms*

The outlook of this study of dispositions has made realist assumptions at various places but this realism is tempered with some reservations. Crispin Wright says that realism consists in a combination of a modest claim and a presumptuous one.[1] The modest claim is that there is a subject-independent reality; the presumptuous claim is that we are capable of describing that reality accurately. I have been leaning on the side of modesty in developing my theory of dispositions. While I have been allowing the existence of mind-independent states of affairs I have also been warning against overconfidence in our abilities to describe these. I have been trying to guard against taking the dispositional–categorical distinction to be anything more than a distinction in the way we talk about instantiated properties or states in the world. The danger is projection of this distinction onto the world such that it is taken to be a division in reality rather than just a division in ways of talking about reality.

The assumption of this modest form of realism leads to an ontology that can be referred to as a neutral monism: monist because dispositional and categorical tokens can be identical, neutral because it refrains from the classification of reality as either 'really' categorical or 'really' dispositional.

The neutral monist says that both categorical and dispositional monism make the mistake of speaking as if their favoured idiom is the one that accurately captures reality rather than the dispositional and categorical idioms being two distinct ways of characterizing the same non-linguistic world. The error of the dispositional and categorical monists' positions could be expressed in the terms

[1] *Realism, Meaning and Truth*, 1.

favoured by Carnap of mistakenly using statements in the material mode rather than the formal mode.[2] When we say that all the world can be characterized in terms of dispositions we are not thereby saying that all the world consists exclusively of dispositional, non-categorical properties. Rather, we are saying that the criterion for application of those terms to the world is met. This does not exclude terms from an alternative idiom being also applicable. The statements:

(1) 'all properties are dispositional'

and

(2) 'all properties are categorical'

are more modestly stated in the formal mode as:

(1a) the concept 'disposition' has universal application to properties

and

(2a) the concept 'categorical' has universal application to properties.

The statement:

(3) 'not all properties are categorical'

should be taken to mean:

(3a) the concept 'categorical' does not have universal application to properties.

(1a) and (2a) have no appearance of inconsistency as (1) and (2) may, hence it is quite possible that all properties could be describable using disposition predicates *and* be describable using categorical predicates.

This kind of position I take to be endorsed in Mackie's remarks on disposition terms that:

The most obvious contrasts are not between properties considered ontologically, as what is there, but between ways of describing them, between modes of introduction, and between kinds of knowledge that we may have about them. We might describe the difference between an iron bar at a very low temperature and the same bar at an ordinary temperature by saying—

[2] R. Carnap, *Die Physikalische Sprache als Universalsprache der Wissenschaft* (*The Unity of Science* (London, 1932)).

if only we knew this—how the molecules were arranged and were moving. Or we might describe this difference by saying that at the low temperature the bar is such that if it is bent, it breaks. The latter is a dispositional description, the former a non-dispositional one. The latter description points to the same difference that is explicitly described by the former description, but it points to it indirectly, by way of an effect that this difference would produce if the bar were bent. Similarly, we may know the difference in question merely as the bar's being such that if it is bent, it breaks. We then know it only as a disposition whereas we know it as a non-dispositional property if we know how the molecules are arranged and move. Now it may well be that most properties are in this sense known only as dispositions, and are therefore unavoidably described and introduced in the dispositional style. But this would not make *what is there* dispositional, and it would not give the concept of a disposition or power any ontological or metaphysical role.[3]

I think that I develop this insight with more consistency than Mackie does. At times it appears that he is failing to apply the same principle to the non-dispositional. Unlike Mackie, I would follow this up by saying that widespread categorical property ascriptions do not make *what there is* non-dispositional.

There are two senses of 'ontology' that are relevant when we are considering the constitution of the world: the first concerns how the world really is and the second consists in how we think the world is. Concepts may determine our ontology in this second sense but obviously leave ontology of the first kind intact. I am thus accepting a division between the world and our conceptualizing about the world. Whatever the world is actually like is unaffected by the way we conceptualize, describe, and think about the world.

This picture, I take it, is consistent with Fetzer's Principle of relativistic realism,[4] which I formulate thus: there is more than one language in which to describe the world though the world exists independently of such languages. Thus, although Fetzer provides a dispositional ontology, embodied in the following claims:

(i) every structural property is a dispositional one,
(ii) every object is a specifically ordered set of dispositions,
(iii) every event is the manifestation of a disposition,

[3] *Truth, Probability and Paradox*, 136.

[4] J. Fetzer, 'A World of Dispositions', in R. Tuomela (ed.), *Dispositions*, 164.

and although he concludes that 'a world of dispositions is a world enough',[5] such an ontology is not necessarily a commitment to the dispositional reductionism I characterized earlier; it is an ontology in the second sense, hence other ontologies in this sense are consistent with it. Fetzer's position could be expressed in the formal mode statement ((1a), above) that the concept 'disposition' has universal application to properties, which is consistent with statements (2a) or (3a).

As regards ontology in the first, realist sense, I am saying that a definite answer can be provided to the quantitative question of ontology, concerning how many kinds of property there are, while we must remain neutral on the qualitative question of whether such properties are dispositional or categorical. I am arguing that in order to be consistent in our claims about causation we can allow only one type of property or state to inhabit the world but whether such properties are dispositional or categorical is a question we cannot answer. All we can say is whether dispositional and categorical concepts are applicable to them.

9.2 *The Place of the Functionalist Theory*

The main contention that arises in the context of these assumptions is that there *is* a difference between the dispositional and categorical forms of discourse and this difference is essentially a difference between functional and non-functional ways of talking about a thing's or a kind's qualities. Clarifying the nature of function-talk is the purpose of this chapter. The functionalist theory I defend draws upon a number of recent developments in the debate that indicate where a correct account of the dispositional–categorical distinction lies. Other important developments have occurred in the philosophy of mind and although these are highly relevant to the solution of the problem of dispositions, those currently working on the problem seem largely ignorant of their importance. I attempt to provide a functionalist theory that can accommodate all the data of our intuitions about dispositions but without any of the counter-intuitive consequences. Although the connection between dispositions and functionalism has been noted previously,

[5] Ibid. 184.

this account will be one that avoids the consequences that were thought to be inevitable corollaries of such a connection. The functionalism that is best suited to a plausible account of dispositions, I will contend, is a homuncular functionalism coupled with a thesis of continuity in levels of nature.[6] The account takes it that nature is composed of a hierarchy of dispositional and sub-dispositional levels where at each level there are structural, or non-dispositional, property instances or states to be identified with dispositions. Further, whether any particular state or property instance *P* is a dispositional property will be shown to be contingent on the description of *P* and the relation in which *P* stands to other states or properties that explain the presence of *P* or are explained themselves by *P*. Nevertheless, in spite of the thesis of the description-relativity of the dispositional, I argue that it still makes sense to call this a realist theory of dispositions, for dispositions construed functionally will be as real as any other properties. Whether any property instance *P* is non-dispositional will also be relative to the description of that property.

Various things can be soluble but what it is in virtue of which they are soluble is that they possess a property that has the functional role of causing dissolution upon immersion in water. What this property's non-functional specification is is left unstated.[7] In cases such as solubility, elasticity, fragility, and the like, the functional role is a causal role specified in terms of typical causal antecedent(s) and typical causal consequence(s). Any property that is known a priori to causally mediate the typical responses to the typical stimuli has the disposition associated with that functional role. It is not merely a matter of fact that it is the soluble things that dissolve when in liquid or the elastic things that can be stretched, though it is an empirical fact which particular categorical bases also occupy those causal roles.

It is not just properties that can have functional essences. Some objects are the objects they are in virtue of the function they

[6] The functionalist account of dispositions presented here draws much from W. G. Lycan's 1987 account of the function–occupant distinction. See 'The Continuity of Levels of Nature', in W. G. Lycan (ed.), *Mind and Cognition* (Oxford, 1990).

[7] Hence Lange, 'Dispositions and Scientific Explanation', is correct to say that different categorical properties could occupy this functional role but I do not see the justification for saying that different categorical properties *must* occupy the role if it is to be a genuine disposition.

perform. One example of such a thing is a thermostat. Something is a thermostat in virtue of it having the function of triggering a switch when a pre-calibrated temperature threshold is passed either from a higher or lower temperature. Any object that has this functional role is a thermostat no matter what its other properties are that realize this functional role. To call something a thermostat, therefore, is to use a dispositional term. All sorts of objects similarly can be cited as having functional essences. 'Thermometer', 'light switch', 'computer', 'bookcase', 'towel rail', 'engine', and 'door handle' can all reasonably be called dispositional terms. The account I give should also be extendable to these cases.

9.3 *Varieties of Functionalism*

Until this point, use has been made of a bare and undeveloped statement of functionalism about dispositions with minimal indication of how this is to be understood. The claim that a disposition ascription is a functional characterization of a state or property instance is far from a transparent or unambiguous claim, as is made clear by the differing varieties of functionalism that have emerged since Putnam (re-)introduced functionalism to the philosophical arena.[8] These various strands of functionalism are a product of the different responses that are possible to a number of issues.

A useful approach is to begin with a basic statement of functionalism to which can be added various refinements in response to the issues that have stimulated controversy. For a basic statement of functionalism about dispositions I take the following two-part statement which specifies, first, in virtue of what a property or state is to be regarded as dispositional and, second, in virtue of what a disposition is the particular disposition that it is.

Basic Statement

(i) What it is that makes a property or state d of an object x a dispositional property or state is that it is a conceptual truth

[8] H. Putnam, 'Minds and Machines', and 'The Nature of Mental States', both in *Mind, Language and Reality: Philosophical Papers*, ii, are particularly significant.

that *d* causally mediates from stimulus events (for which *x* is patient) to manifestation events (for which *x* is agent).

(ii) What it is that makes a disposition *d* the type of disposition it is is the specific stimulus and manifestation events to which it bears the relation of causal mediation.

A few words of clarification are necessary. In part (i) of the basic statement the stimulus and manifestation events are distinguished as events for which *x* is patient and events for which *x* is agent. By including these stipulations I am trying to make a qualitative distinction, explicated in terms of relation to *x*, between the events classed as stimuli, which are events which exert a causal influence upon *x*, and the events classed as manifestations, which are events which *x* exerts its causal influence upon (in ideal conditions and given the stimulus event) in virtue of the mediation of *d*. The agent–patient distinction need not be taken too seriously. Arguably these are nothing more than convenient ways of understanding events and their relation to objects in the world. As such, there seems little justification for understanding them to have serious ontological import.

Part (i) of the basic statement is the claim that *d* is a dispositional property iff *d* is a functional role occupying property by definition. Note that this final clause, that the functional role be occupied *by definition*, is essential if any non-dispositional property terms are to be distinguished. In accordance with the causal criterion of property existence and the argument from identity of causal role, all properties will play a variety of functions in a variety of situations. What counts is whether the connection to a particular causal role is definitional and, as we will see, this is where the relevance of modern teleological accounts of functionalism will come in.

Part (ii) of the basic statement specifies how different individual dispositions are distinguished. Particular pairs of stimulus and manifestation events can determine different dispositions for the properties that have the appropriate relation of causal mediation between them. A corollary is that any property or state *d′* that has the appropriate relation of causal mediation between any stimulus and manifestation pair $<s', m'>$ is a specific disposition even if there exists no concept corresponding to *d′* with which we could conceptualize about that disposition.

Recent development of functionalism in the philosophy of mind has shown that there are a number of issues concerning the interpretation of the basic statement. The aim is to find the kind of functionalism that is required specifically for a plausible theory of dispositions. This does not mean that the same brand of functionalism is also being offered as the most suitable version for the philosophy of mind. There is no reason to expect one treatment to be fitted for both subject matters. There are certain problems for one case that are not considerations for the other. Plausible theories of the mind have to contend with phenomenal qualities, for instance. They have to account for the way experiences can 'feel' to someone with a mind and it is suggested as a notorious failure of functionalism in the philosophy of mind that such experiences cannot be functionally captured. Arguably, some state may have all the appropriate causal relations to inputs, outputs, and other states and yet still not be pain in virtue of the simple fact of not feeling painful to its owner. Alternatively there may be a painful feeling which, for any number of reasons, does not hold the appropriate relations to inputs, outputs, and other states.[9] This special problem of phenomenal qualities is one which has no analogue to consider for the functionalist about dispositions. The area of dispositions brings its own special problems. The functionalist about dispositions has to find a version of functionalism which is consistent with attractive accounts of the dispositional–categorical distinction, the causal role of dispositions and the world relativity of disposition ascription. Fitting the theory to these requirements means that there is no reason to expect a single brand of functionalism to be the best for both dispositions and the mind. Each case must be considered independently.

The best functionalist theory for dispositions is produced by appropriate responses to the following debates: the relationship between functions and dispositions; the causal role of functional states; functional specification versus the functional state identity thesis; teleology and the relativity of disposition ascription; homunctionalism and continuity in the levels in nature. I will consider each of these issues in turn and describe how they impact on the theory.

[9] Such problems are considered by D. Lewis, 'Mad Pain and Martian Pain', in N. Block (ed.), *Readings in Philosophy of Psychology*, i (Methuen, 1980).

9.4 *Functions and Dispositions*

An important issue needs to be considered to make way for any functionalist theory of dispositions. This is the question of the relationship between functions and dispositions. Some basic questions need answering. Are all ascriptions of functions ascriptions of dispositions? Are all disposition ascriptions functional ascriptions? The latter is true for any functionalist about dispositions but if they also say that all functions are dispositions, then the concepts of function and disposition become equivalent and are collapsed into each other.

This is a possible source of objection that is made particularly acute by the work of Robert Cummins.[10] Cummins has suggested that unless it is detailed what exactly is meant by a functionalist understanding of dispositions, then there is a danger that little will be gained from the analysis. What is the point in saying that a disposition ascription is a functional characterization of a property if to say that something has a function is just to say that something has a disposition? Cummins thinks that function ascriptions entail disposition ascriptions so the budding functionalist has to be aware of the danger that any attempt to analyse disposition ascriptions functionally will result in circularity.

Another slant on this problem is that if a functional role has to be analysed in terms of conditionals, then it is evident that nothing has been gained because the disposition ascription itself would have been given the same analysis as the functional one. Thus shifting the focus of attention from dispositions to functions produces no benefit at all. If Cummins has the correct analysis of function ascriptions, then any functionalist theory of dispositions is a trivial non-starter.

However, setting aside the point that, if true, it could well only be an example of the paradox of analysis, I think a strong case can be made for saying that Cummins has got the relationship between functions and dispositions the wrong way round and that, understood correctly, to say that disposition ascriptions are functional ascriptions is informative. Consider Cummins's basic claim:

[10] R. Cummins, 'Functional Analysis', *Journal of Philosophy*, 72 (1975). Cummins has made the objection more explicit in correspondence.

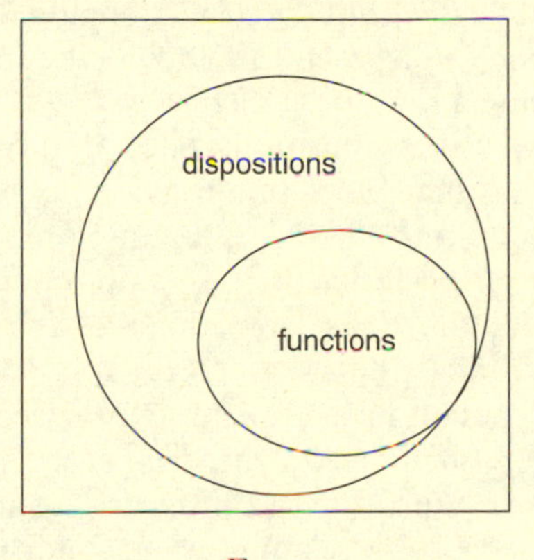

FIG. 1

Something may be capable of pumping even though it does not function as a pump (ever) and even though pumping is not its function. On the other hand, if something functions as a pump in a system *s* or if the function of something in a system *s* is to pump, then it must be capable of pumping in *s*. Thus, function ascribing statements imply disposition statements; to attribute a function to something is, in part, to attribute a disposition to it. If the function of *x* in *s* is to *F*, then *x* has the disposition to *F* in *s*.[11]

Cummins's claim here is that something can have a capacity to *F* though it does not have the function to *F* but, in contrast, if anything has the function to *F*, then it does have the disposition to *F*. If we treat dispositions and capacities equivalently,[12] then Cummins's envisaged relationship between dispositions and capacities is as in Fig. 1.

However, the sense of function being used by Cummins seems to be an atypical one which is probably not the sense intended in

[11] Ibid. 185.

[12] Cummins thinks that a distinction can be drawn between dispositions and capacities that I do not. His basis for the distinction comes entirely from ordinary usage. He cites as evidence, for instance, the feeling that (1) 'hearts are disposed to pump' feels strained, whereas (2) 'hearts are capable of pumping' does not. If we are to follow ordinary usage, then I think there is a lot to be said for the Rylean view that a capability is a disposition it is useful to have, an incapability is the lack of a disposition it would be useful to have. On this view, capacities, capabilities, tendencies, and so on, are all dispositions with something else attached. A 'capability', for instance, is a disposition accompanied by a value judgement as to the usefulness of that disposition. I see nothing important as to ontology coming from this point so I think it is a distinction of no importance for the present issue.

most theories that call themselves functionalist. Cummins's claim that something can be disposed to *F* though it does not function to *F* seems to be using a sense of 'function' where *x* has a function to *F* iff *x* is actually used at some point to *F*. This appears to be what is meant by saying that 'Something may be capable of pumping even though it does not function as a pump (ever).' The proposed contrast between dispositions and functions consists in the possibility that, unlike functions, something can be disposed to *F* though it never in fact *F*'s.

The sense of function in the functionalism recommended in the present and other functionalist theories is one that accepts that something can have a function to *F*, in certain conditions, though it never actually does *F* because appropriate conditions for *F*-ing are not realized. The function of a can-opener is still to open cans even if it never actually opens a can: for instance, if all cans are destroyed. This means that the case Cummins recommends as a possibility, where something has a disposition but not a function to *F*, would not be supported on the grounds Cummins gives. Clearly there is a reasonable sense of 'function' that corresponds to the claim Cummins is making but it is not exclusively the kind of function in which functionalists are interested.

The response of the functionalist about dispositions to Cummins's attack should thus be along the following lines. Cummins is using a sense of 'function' where a function-ascribing statement is understood as a causal-role ascribing statement; that is, the ascription of what I have called a concrete disposition. This is too narrow a sense of 'function' because not all function ascriptions are ascriptions of concrete dispositions or ascriptions of dispositions of any kind. I have made the claim that some dispositions may be understood as abstract, such as being divisible by 2. A functionalist theory of dispositions has the advantage of being able to explain why it is plausible that such abstract powers are dispositional in addition to the more commonly cited concrete dispositions of fragility, solubility, and the like. If a theory of dispositions can include such cases, at no added cost, then there seems no objection to including them even if there may be certain grounds upon which abstract dispositions are atypical.

However, a second and crucial point is that not all true function ascriptions are disposition ascriptions. A disposition is only one particular kind of function. Thus, in contrast to Cummins, I am

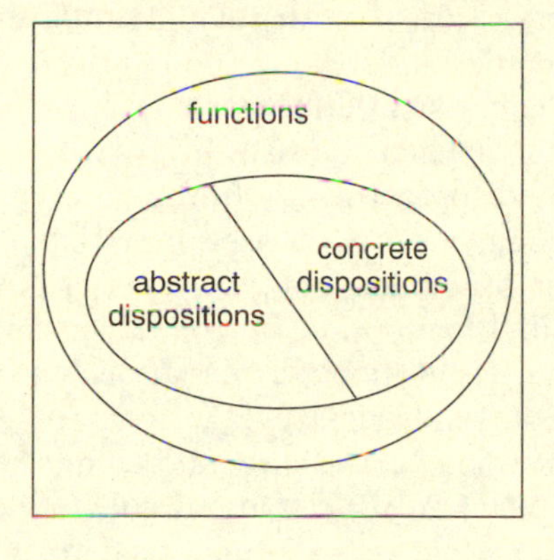

FIG. 2

offering the account represented in Fig. 2. On this view, things can have functions but not dispositions. Examples could be the function of a flag being to add grandeur or the function played by a premiss in an argument.[13] There are legal functions played by a magistrate, the function of a protest, the function of a rule in a game, the function of a road sign. These are fairly commonplace cases where in saying that something has a function we are not saying that it has a disposition to do something. There are functions of things that are determined by *convention* such as flags being a symbol of grandeur or a road sign being an instruction. These functions depend essentially on our responses to certain objects or symbols and for this reason it would seem a mistake to ascribe a disposition to such an object. The disposition cannot be ascribed to the object completely because of the response-dependency of the function in question which means that it is a function that can be held only relative to minds. There are functions that are determined by *relations* to other items in a system such as rules of a game or the function of a premiss in an argument. These functions would make no sense and would not be possessed outside the system. A premiss which is part of no argument is no premiss at all but only a proposition. Dispositions, in

[13] Whether a proposition is also a premiss, or indeed a conclusion, is dependent upon the function it plays in an argument.

contrast, are possessed in the absence of conditions which would provoke their manifestation.

The charge originating in Cummins, that it is of no benefit to the understanding to say that disposition ascriptions are functional characterizations of properties is shown, through these examples of non-dispositional functions, to be based on an assumed relationship between disposition and function ascription that has little to recommend it. By adopting functionalism about dispositions we are saying that dispositional explanation is a variety of a more wide-ranging, not less wide-ranging, form of explanation and characterization where something is being said about a role that can be played. For our paradigm concrete dispositions this role is a causal one but not all functional roles are causal, nor even dispositional.

9.5 *Functionalism and Causation*

Problems associated with the causal role of dispositions have been considered in depth in Chapter 6 but there remains another problem that arises specifically from a functionalist understanding of dispositions. The causal status of dispositions is threatened by an understanding of functional properties that is urged by Prior and was considered briefly in Sect. 8.5. This is the view that functional properties are second-order properties. How this view threatens the causal status of dispositions is clearly articulated by Block. Discussing functionalism in philosophy of mind, Block says:

> [F]unctionalism dictates that mental properties are second-order properties that consist in having other properties that have certain relations to one another. But there is at least a *prima facie* problem about how such second-order properties could be causal and explanatory in a way appropriate to the mental. Consider, for example, provocativeness, the second-order property that consists in having some first-order property (say redness) that causes bulls to be angry. The cape's redness provokes the bull, but does the cape's provocativeness provoke the bull?[14]

Given that provocativeness has the appearance of being a dispositional property, then the question whether it is first-order redness

[14] N. Block, 'Functionalism', in S. Guttenplan (ed.), *A Companion to the Philosophy of Mind* (Oxford, 1994), 331.

that causes the bull to be angry or second-order provocativeness appears to be the very same problem as whether it is a first-order molecular structure that causes sugar to dissolve or its second-order solubility. If a functional property is a property of having a property, as Prior understands dispositions, then it seems that they can add no extra causal powers to their possessors over and above those causal powers of the first-order properties that realize them.

I have already indicated that I think this is a mistaken view of the relation between dispositions and their bases. I am willing to offer an alternative account for the case of solubility but not for that of provocativeness which I think is an example that will mislead us if we try to generalize from it because it should be classed among judgement- or response-dependent dispositions. Because of this extra element of response-dependence there is no prospect of provocativeness being understood as something more than having a property with the causal role that red things have in actuality. The colour which provoked anger may have been different because which colour does this depends on the judgements, rational or non-rational, of the observing subject. The solubility of a substance I take to be a non-judgement-dependent disposition concept, however, so there is plausibility in understanding such a property as second-order: the property being the possession of a particular type of molecular structure. How can the functionalist preserve the efficacy of dispositions in this case?

The answer is Davidsonian. Given the identity of the tokens that instantiate dispositions and their causal bases and given that causal bases are causally efficacious of disposition manifestations, then dispositions are also causally efficacious of their manifestations. There is, admittedly, a trivially analytic connection between having a disposition and manifestations of a certain kind in certain circumstances but this does not prevent the disposition from being a cause because that same state can be redescribed in a way that is not trivially connected.[15]

If we are to take the identity theory and functionalism seriously, then it is a mistake to describe dispositions as second-order properties of having properties with a particular causal role. A move to the language of states of an object or substance more clearly illustrates this. The argument for identity dictates that the disposition actually

[15] See Sect. 6.8.

is this very same state which has role *R*. Therefore, although to have a disposition is to have a state that plays causal role *R*, having a categorical base is also having a state that plays role *R*. The only difference between the two types of state ascription being that one and not the other has a conceptual, a priori connection to *R*.

9.6 *Functional Specification versus Functional State Identity Theory*

Block discusses the difference between the functional state identity thesis (FSIT) and functional specification theories of functionalism.[16] The difference between the two views is as follows.

Functional specification views are in Armstrong and Lewis.[17] The functionalism consists in the claim that a mental state, such as pain, is really a physical state but one which is functionally specified. This functionalism about mental states is supposed to support physicalism because pain and any other mental state is actually a physical thing, c-fibre firing in the case of pain perhaps, though it is a physical thing under a functional rather than explicitly physical description.

FSIT, from Putnam and Fodor, is anti-physicalist.[18] Mental states can be analysed as functional states only and these may be realized, in theory, by something non-physical. Thus, even if all mental states are actually realized by physical states, reductionism is still not true because mental states can still be said to be identical only to functional states not to physical states.

How should the functionalist about dispositions respond to this debate? Clearly the analogous question has to be asked. It would be formulated in the question: is a disposition a functional state which could, logically, be realized by any kind of state at all or is a disposition a functional specification of a categorical state? Given what has been said so far, particularly the endorsement of property monism and Lewis's argument from identity of causal role, it may

[16] Block, 'What is Functionalism?', in N. Block (ed.), *Readings in Philosophy of Psychology*, i, 177–81.

[17] D. M. Armstrong, 'The Nature of Mind', in N. Block (ed.), *Readings in Philosophy of Psychology*, i, and Lewis, 'Psychophysical and Theoretical Identities'.

[18] H. Putnam, 'The Mental Life of Some Machines', and J. Fodor, 'Explanations in Psychology', in M. Black (ed.), *Philosophy in America* (London, 1965).

be thought that functional specification is the most natural position to support. After all, aren't all dispositions just said to be categorical terms under a functional specification? In actual fact, the functionalist theory of dispositions defended is a version of the functional state identity theory.

As Block suggests, some of the disagreement between functional specification and FSIT comes down to the fact that functional specifiers are actually type–type identity theorists.[19] Lewis goes to elaborate lengths to defend type–type identity, for instance.[20] If pain actually is type–type identical with c-fibre firing, then it has greater plausibility to say that pain is c-fibre firing functionally specified. However, the view of dispositions I recommend is one where we would not say that a disposition D is a categorical state C functionally specified even if D and C are type-identical. To say this would be to misunderstand the meaning of a disposition ascription and the force of the argument from identity of causal role.

A disposition ascription is a functional characterization of a property. It is to say what causal contribution the possession of that property makes to its owner without saying anything about how that causal role is brought about. These causal roles are actually occupied by categorical bases but we cannot straightforwardly reduce dispositions to such bases. The identities between bases and dispositions is not an analytic matter. Occult forces and magical states, if they existed, could also occupy these causal roles. The argument from causal role, however, leans on the empirical premiss that dispositions actually do have these categorical bases. From analysis of a disposition alone, we can say only that it is a functional state. Only by reliance on the empirical premiss can we say that as a matter of fact these functional roles are realized by categorical bases so we cannot say that a disposition ascription is an ascription of a categorical state functionally specified.

9.7 *Relativity of Ascription and Teleology*

One consequence of this view is that a property or state ascription is dispositional (or categorical) only relative to a choice of *explanandum*. This is a point particularly suited to a functionalist theory

[19] 'What is Functionalism?', 180.

[20] 'Mad Pain and Martian Pain.'

of dispositions because it is a point that has been developed in the recent teleological interpretations of functionalism.[21] Taken together, the insights available from this teleological functionalism about dispositions provide another path to a neutral monist ontology.

In explanation of this point it needs to be specified in what sense a state ascription's dispositional status is a relative matter, the relative conditions in which a predicate is dispositional and those in which it is not.

First, what is the dispositionality of a term relative to? We have seen, according to the position argued in Chapter 7, that the same property or state *S* of an object or substance *x* can be denoted by non-synonymous predicates. Whether or not *S* is understood as a disposition or a categorical state of *x* is relative to the predicate '*P*' used to denote it. Furthermore, whether '*P*' is a dispositional or non-dispositional predicate is dependent upon it having the correct relation to a pair of events which are possible stimulus and manifestation events for dispositions.

The dispositional–categorical distinction can then be stated in a simplified form that whether *P* is a dispositional or a categorical predicate is determined by whether '*P*' is a functional or non-functional characterization of *S*. However, recent teleological interpretations of functionalism have begun by pointing out that it is not obvious what it is for something to be a function for any particular state, artefact, or organ. What I have done so far makes the distinction in terms of the relation of a property or state-denoting term to a particular causal role. What was offered as the criterion of the dispositional was whether a predicate term was one that contained within it, as a conceptual truth, the notion of an appropriate relation of causal mediation between a particular stimulus event and a particular manifestation event. But in what sense can the predicate term contain within it the appropriate relation between the appropriate events? According to teleological interpretations of functionalism it does so because, like all functional characterizations, disposition terms set up a distinction

[21] See L. Wright, 'Functions', *Philosophical Review*, 82 (1973); R. Cummins, 'Functional Analysis'; and E. Sober, 'Putting the Function Back into Functionalism', in Lycan (ed.), *Mind and Cognition*.

between behaviour which is 'accidental' and behaviour performed as a function of a thing (state, artefact, or organ).

A thing can be responsible for many different types of behaviour. A state may be responsible for dissolving in liquid but also for tasting a particular way. The heart may be able to pump blood but it also makes a noise. A telephone may facilitate audio communication but it also absorbs and reflects light. In the last two examples, Larry Wright notes that we think that one type of behaviour for these things is its function; the other type of behaviour is something it does just by 'accident'. The functionalist needs to be able to state some principled reason why some types of behaviour seem to be functions of things and yet other types of behaviour seem only accidents. Wright thinks that the way the functionalist should make this distinction is in terms of teleology. This is not intended as introducing any occult or animistic powers to objects such as telephones. The teleology is understood in the following sense. Facilitating audio communication is the function of a telephone, rather than absorbing and reflecting light, insofar as it is because of its facilitation of such communication that the telephone is there. It is because of this ability, among all the other types of behaviour of which such an object is capable, that telephones are manufactured and installed in homes and offices. In the case of functions of organs and organisms the notion can be explicated in terms of evolution. It is because of the heart's blood-pumping behaviour that it was naturally selected for certain organisms rather than its noise-making behaviour. We have hearts because they pump blood, not because they make noise, hence to pump is the function of the heart rather than noise-making.

How does the distinction between function and accident apply in the case of dispositions? Consider the plausible example of a state which facilitates both dissolving in water and tasting a certain way on the tongue. Which of these two types of behaviour is the function of that state? The answer is that what behaviour is the function of the state is relative to which of the possible predicate terms that can denote that state is being used. The predicate 'solubility' may be used to denote the state in which case dissolving in water would be understood as the function of the state and tasting sweet would be an accident of it.

That dissolving in water is the function of solubility needs to be given some justification slightly adapted from Wright's account because it seems to make little sense to say that it is because of dissolving in water that solubility *is there*. Some meaningful interpretation of this is not far away, though. We can say that it is in virtue of causally mediating dissolving upon immersion in water that something is soluble whereas it is not in virtue of tasting sweet to the tongue that it is soluble, even if the very same state is responsible for both forms of behaviour. However, relative to some other predicate, tasting sweet on the tongue can be understood as the function of that state: where this is a predicate such as 'sweet' that can be used for that causal role. For some other predicate which denotes that same state, perhaps some molecular-structural predicate, none of these behaviours is its function, as a matter of conceptual necessity, though it does facilitate many types of behaviour accidentally. A predicate which is conceptually connected to no function, though is connected to behaviour accidentally, is a categorical predicate and here we have another way of understanding the dispositional–categorical distinction.

This gives us a better understanding of how a single state can be denoted by non-synonymous predicates and how some of these can be dispositional predicates and others categorical. We also have a better understanding of what it is for a characterization of a state, artefact, or organ to be a functional one. This supports the neutral monistic understanding of dispositions in that it shows how the dispositionality or non-dispositionality of a property or state is not a question grounded in ontology but, rather, a matter relative to the predicate with which the property or state is denoted and the connection between this predicate and behaviour which is mediated.

Now that the relativity of disposition ascription has been introduced, the functionalist theory of dispositions can also be shown to fit with two other recent innovations in the theory of functionalism, namely, homunctionalism and continuity in the levels in nature.

9.8 *Homunctionalism*

Homuncular functionalism or homunctionalism is said by Lycan to have originated in Attneave though it is developed by Dennett before being offered as a full part of functionalism in Lycan's own explicit formulation.[22]

The notion of a homunculus, rather like that of teleology, is one that has come in for some knocks. However, understood in the way developed since Attneave it has emerged as a respectable and useful concept on which the philosopher can draw.

Homunculi were disparaged where they were irredeemable causal powers standing as stop-gaps where a genuine explanation was lacking. The classic homunculus occurs in the philosophy of psychology where a mental ability may be provided with a putative explanation in terms of some component of the mind being endowed with the very same ability that it is supposed to explain. This is no explanation at all because the possession of the same ability at the lower level is itself unexplained. The question can always be asked of how the homunculus in the head can do what it does and the same strategy applied to the homunculus means that the strategy initiates a regress.

Modern homuncular functionalism avoids this charge by not empowering the homunculus with the same ability that, at the higher level, is to be explained. The lower level homunculi are more stupid. They have less sophisticated abilities which are possessed in virtue of even lower-level homunculi with even less sophisticated abilities. These descending levels come to an end with homunculi with the most basic abilities consisting in simple mechanisms requiring no intelligence at all. The homuncular strategy is, therefore, genuinely explanatory rather than regressive.

How can this strategy be used in the case of a functionalist theory of dispositions? The view encouraged is as follows. A disposition of an object can be explained in terms of its component parts. The explanation works by itemizing what these component parts do: what contribution they make to the behaviour of the whole. In other words, the explanation of the dispositions of the whole is in terms of the dispositions of its parts where each component part is understood functionally according to the abilities it can bestow. These components will be more numerous and will have abilities which are simpler and make only a partial

[22] See Lycan, 'The Continuity of Levels of Nature', 79. His sources are F. Attneave, 'In Defense of Homunculi', in W. Rosenblith (ed.), *Sensory Communication* (Cambridge, Mass., 1960), and D. C. Dennett, 'Why the Law of Effect Will Not Go Away', in Lycan (ed.), *Mind and Cognition*.

contribution to the disposition of the whole. Hence this is not the empty explanation of the classic homunculus.

The sub-dispositions of the components are, in turn, possessed in virtue of the sub-sub-dispositions of their own parts so that at each level of explanation a further explanation is available, in terms of structural components, with which to redeem the homuncular explanation of the level above. The explanation goes bottom-up as well as top-down. The original disposition we began with may itself go on to be a component of a larger system. The notions of whole and component are themselves relative.

As an illustration of a homunctionally explained system, take the disposition of a car engine to afford propulsion. A mechanic could offer us an initial explanation of how the engine works. They might make reference to components such as carburettor, distributor, radiator, throttle, alternator, and so on. These components, at this stage, are being functionally described. For these purposes, they are essentially things with dispositions: things that perform certain mechanical functions. Further explanation could, of course, be provided by the knowledgeable mechanic. The disposition of the whole, in this case the engine, has been explained in terms of the dispositions of its parts. These parts can in turn be treated as wholes whose behaviour can in turn be explained by the same strategy. They have their dispositions in virtue of their parts, and so on.

The engine is itself one component of the whole vehicle. Its abilities can be explained in terms of the dispositions provided by the engine but other components will need to be included, such as wheels, steering, and brakes, if the workings of the whole are to be understood. A key notion that distinguishes this form of homuncular explanation from its impoverished ancestor is that of corporatism. Homuncular functionalism takes entities to be corporate: possessing dispositions in virtue of a set of sub-dispositions which are more numerous but which work together performing simpler tasks. Out of the combination of these numerous simple tasks comes complex behaviour.

One question that arises, however, is what is at the bottom of all these levels of explanation that supports the whole? Presumably even this form of homunctional explanation cannot go on forever. If the explanation could never be completed, then even this kind of explanation is incomplete.

Incomplete explanations are tolerated in practice for dispositions of the tiniest parts may have to be explained in terms of the dispositions of molecules, which are explained in terms of the dispositions of atoms and then subatomic explanations. It is an immodest claim that all the workings of these levels are understood and there are often points where it is wise to say that this is as far as we understand. However, the reliance upon some unexplained dispositions is tolerable. Such dispositions would have to be simple and have no structural components which can be accredited responsibility. These are precisely the sorts of things that I have discussed earlier which we could refer to as ultimate or ungrounded dispositions. According to some interpretations, the whole of physics rests upon the foundation of such assumed ultimate dispositions, as I will discuss in the next chapter.

9.9 *Continuity in Levels of Nature*

Homunctionalism leads us to a continuous picture of reality. This notion is developed by Lycan who takes as his target those views that attempt to draw a sharp distinction between 'software' and 'hardware' in the case of the mind. He calls this kind of division 'two-levelism'. There is a kind of two-levelism that pervades the discussion of dispositions. This occurs when the dispositional–categorical distinction is thought to be an absolute matter applying at the level of ontology. The view that has replaced it is one where it is a relative matter whether something is a disposition or a categorical base and if this view is plausible then positions like the property dualism I attribute to Prior, Pargetter, and Jackson face a further difficulty: it is not a truth of ontology whether a property is dispositional or not; it can only be a truth about concepts.

Contrary to two-levelism, where we are asked to accept a clear bifurcation of reality, the continuity picture applied to functionalism about dispositions takes it that there is a continuous hierarchy of levels in nature with both dispositional and categorical to be found at all levels. According to the argument from identity of causal role, these dispositions and categorical bases at the various levels stand in identity relations. The distinction we draw in the world between 'function' and 'realizing stuff', which the distinction between dispositional and categorical exemplifies, is not absolute at all. With the hierarchy view the function–structure distinction 'goes relative' in that something is a structure as opposed to a function only relative to a particular perspective.[23] A particular molecular structure may be the occupant of a functional role relative to a higher level of nature, and a function relative to a lower level of nature.

Displaying close similarity with the present functionalist theory of dispositions, Lycan opts for identity between the mental (functional) and the homunctional (occupant of functional role). He says: 'I propose to type-identify a mental state with the property of having such-and-such an institutionally characterized state of affairs obtaining in one (or more) of one's appropriate homunctional departments or subagencies.'[24]

A further feature is that either end of the continuum may contain states of affairs that are characterized only in terms of one class—wholly dispositional or wholly categorical—but this fails to establish categorical or dispositional monism, for a different characterization may be forthcoming in time, and the vast majority of states of affairs are describable in neither class of terms exclusively.

Lycan's picture, if accurate, is one that embraces the diversity of description afforded by the dispositional and categorical idioms. Both have their specific purpose, and this is a purpose which could not be fulfilled by the other category. A world described entirely in categorical terms is a static world and hence, not ours. A full description of our world must be one that permits change but a description that was entirely in terms of structure contains no indication of the changes that occur in the world. This merely follows from the familiar Humean point that it is a logically contingent matter what effect follows from any antecedent event and one suggestion is that laws of nature can be added which animate the world. An alternative strategy is to ascribe dispositions to things and I will assess these two approaches in the next chapter.

The functionalist theory of dispositions states, to summarize, that nature consists of hierarchically ordered levels where at each level there are entities or properties that can be characterized either dispositionally or structurally and whether they are dispositional or structural characterizations depends on their explanatory relations to other entities or properties within the whole of which

[23] 'Continuity of Levels of Nature', 78. [24] Ibid. 81.

they are a part. Generally, levels of entities/properties are taken to be non-dispositional when they are components of some higher level disposition. This is consistent with the dispositional–categorical distinction outlined in Chapter 4 because the distinction made *P* a dispositional property if and only if *P* entailed certain conditionals as a matter of conceptual necessity. Now given that all properties have causal roles, whether any property *P* is to be construed as a dispositional property or not is a matter relative to a particular causal role. If a causal role is one that *P* occupies as a matter of conceptual necessity, then *P* is dispositional relative to that causal role; if it doesn't occupy that causal role as a matter of conceptual necessity, but does so as a matter of fact, then it is non-dispositional relative to that causal role. Relative to the functional role of causing dissolving when in liquid, a denotation of a property *P* that has that causal role by conceptual necessity—solubility—is dispositional. Denoted in such a way that does not conceptually necessitate that causal role, perhaps in terms of molecular structure, that same property comes out as non-dispositional. By the argument from the identity of causal role, however, these two denotations are of the same state or property instance.

I hope I have demonstrated adequately that the functionalist theory of dispositions is the most plausible account available. I have tried to show that it is consistent with all our pre-theoretic commitments, without counter-intuitive consequences, and that it delivers explanatory advantages to the areas where appeal to dispositional properties typically is made.

10

Laws of Nature Outlawed

10.1 *Two World Views*

In the chapters preceding I have gradually been assembling the components of a theory of dispositions. Commitment to a number of theses has been recommended. They are realism about dispositions *qua* instantiations of properties, a functionalist construal of disposition ascriptions, and an anti-reductionist identity theory for dispositions and their categorical bases. The job of providing a theory of dispositions is done and I have a different purpose in this chapter.

Disposition ascriptions play an explanatory role and dispositions can be conceived of as units of explanation. I have developed an ontology that legitimizes their explanatory status by laying down the foundation of dispositions as real properties, to which explanatory appeal can usefully be made, rather than disposition ascriptions being elliptical expressions for events described in conditionals. I want to do more than just vindicate appeal to dispositions, however. In this chapter I aim to promote positively such appeal. Specifically, I will be considering the case for an ontology of real dispositions replacing the so-called laws of nature as the basic building blocks of explanation. I will argue that they, instead of such laws, should be understood, in certain cases, as the 'brute', fundamental facts which are the basis of all other facts. The ontology for dispositions that I have developed in the preceding nine chapters is fit for supporting such a world view.

To appreciate what is at issue in the decision between laws and dispositions we can consider how we are to explain the phenomenon of change.

Change evidently occurs in the world. A change is an event. Some events are understood ordinarily to be disposition manifestations such as the passing of sugar from a solid to a dissolved state. The class of disposition manifestations also includes the reflection of red light rays from the surface of an object, the production of a note from the plucking of a string on a double bass, and so on.

Could all events be considered to be disposition manifestations? Conceivably all events could be if we had a sustainable ontology that conceived of the world as a conglomeration of dispositions that manifest themselves when prompted.[1] This view has its Humean detractors who offer their own ontology and their own explanation of change. There are two rival world views that each purports to offer an explanation of the way change is produced. It will be allowed that each world view has its initial credibility. How, then, are we to adjudicate between the two? One strategy is to find something that one ontology can explain that its opponent cannot. If I succeed in my aims, I will do just this.

I will call the world views in question the *laws view* and the *dispositionalist view*. The former is characterized by a reliance on general truths interpreted as constant conjunctions or real laws of nature as the ultimate, inexplicable, units of explanation to which we must appeal to account for the events that populate our world. The dispositionalist, on the other hand, claims that some notion of a real power, disposition, or capacity can fulfil the same role that general laws do in the laws view but that particular capacities can fill this role with none of the disadvantages inherent in general laws. The dispositionalist has his or her own problem to face: of providing a plausible account of events in terms of real dispositions that does not itself depend on laws. I will argue that there is good reason to believe that can be done and that, coupled with the problems encountered for laws, we are compelled to adopt the dispositionalist rather than laws world view.

It may appear that the account of dispositions I have developed up to this point is incompatible with an anti-laws conclusion insofar as it has been dependent upon the notion of a law of nature at a number of places. The dispositional–categorical distinction, for example, itself depended upon the contingency of laws of nature. The distinction can be made under the assumption of the laws of nature being contingent but the assumption can be amended. It is convenient to state the distinction in such terms because of the familiarity of the notion of a law of nature being

[1] Fetzer, 'A World of Dispositions'.

logically contingent. I will go on to show, however, that if laws are abandoned, then what takes their place can provide the contingency necessary to support the dispositional–categorical distinction. At no point will the account of dispositions that I defend be dependent upon an ontology of laws for which it is being offered as a replacement.

I will show how both the laws and dispositionalist strategies face difficulties. They are not, however, difficulties of equable value. While I will argue that the problems for laws are considerable and such that we are not well motivated to evade them, we are well motivated to solve the problems for individual capacities for these are such that they can be solved by adopting the right theory for dispositions.

10.2 *Events Explained by Laws of Nature*

It is laws of nature understood, at the very least, as general facts making general statements true that are the target for replacement by the dispositionalist ontology. Hume would not allow a notion of laws as natural necessities but he would allow that there were general truths about contingent, constant conjunctions of events. The view that there is nothing more in nature than such constant conjunctions is one that not all have followed but, I suggest, many such realist opponents of Hume have remained distinctly Humean in their approach. General truths as an explanatory basis have been kept but something has been added as reinforcement. Natural necessity, for instance, has been offered as a truth-maker and guarantor for such general truths. The dispositionalist view opposes all such accounts of general laws, whether they have a minimal Humean construal or are reified as natural necessities. Any such general law account I will take to be part of a Humean world view even though there are evident differences in detail.

The laws view I take, therefore, to be something like this. Events occur. Commitment to events may bring a commitment to properties because a typical event will be:

(e_1) a is square at t_1 but round at $t_1 + \theta$.

Events are, however, the basic ontological unit in this world view and properties are parasitic upon them. Properties exist only in virtue of their instantiations in events. All such properties are categorical in the sense that all that there is to properties is manifest in the present and they contain no hidden possibilities waiting to be unleashed.

Such events come as independent existences among which there are no logically necessary connections, nor are there any between events and properties. Thus, given any particular instantiation of a property and the occurrence of an event, it is entirely a matter of contingency what subsequent event will occur. It follows that there is no logical impossibility, for example, in sugar being immersed in a liquid being followed by it exploding, or turning into a white rabbit, rather than the sort of event that we are accustomed to observing in such situations: its dissolving. This point is nothing to do with the fact that certain conditions may not be right for the dissolving of the sugar—an issue I considered in Sects. 4.7 to 4.9—but that in *any* conditions there is no logical necessitation between events.

Nevertheless, there is some order to the world. In an important sense it *is* necessary that sugar in liquid dissolves but this necessitation is not logical. The traditional view which originated in Hume has it that what it must be that brings order to the world is either some fact about the way the world is—a constant conjunction—or something stronger: natural necessity, which is logically contingent. The laws of natural necessity, therefore, could have been otherwise.

Some of these constant conjunctions or natural necessitations are ultimate truths about the way the world is. These are ultimate facts about the world in virtue of being truths for which there is no further explanation. It is thus unanswerable why the basic laws of nature are as they are and not otherwise.[2] In other words, it is unanswerable why it is a world with *these* laws that is actual instead of one of the many other possible worlds which have different laws.

We describe the conjunctions or necessitations in general statements which are not the laws of nature themselves but, rather, the linguistic articulations of the laws. By adding such general statements of laws to our properties and events we get explanations for those events or changes that so far have been called disposition

[2] I am assuming a naturalistic version of the Humean picture, i.e. I permit, in this context, no appeal to a supernatural explanation of the actual set of laws.

manifestations. Our laws world view thus explains apparent disposition manifestations like so:

[wv_L] (events + laws of nature) $\circ\!\!\rightarrow$ (events describable as disposition manifestations),

where '$\circ\!\!\rightarrow$' represents an as yet undefined relation which could be anything as weak as conjunction, in the original Humean account, or as strong as a relation of natural necessitation.

This kind of explanation affords another opportunity for eliminativism about dispositions for all events can be explained wholly without reference to them. The supporter of the laws view can therefore deny the existence of dispositions and posit only events and laws (plus any properties, where required, which are all categorical). There is no such event as:

(e_2) *a* is soluble at t_1.

Rather, such a disposition ascription is analysable into complexes of events described in conditionals as we saw with the conditional analysis in Chapter 3:

(e_c) *a* is put in water at t_1 $\circ\!\!\rightarrow$ *a* dissolves at $t_1 + \theta$.

The supporter of the laws view faces recalcitrant problems, however. These are problems about the content of such laws, their truth, and the sense in which laws can be said to exist as anything above regularities of actual events. A defence of laws in the face of such problems is a considerable task. Before looking at them more closely, however, I will outline the dispositionalist world view.

10.3 *Events Explained by Dispositions*

The dispositionalist attempts to dispense with laws of nature by grounding natural necessities within the individual instantiated properties themselves. This gives us the model of instantiated properties as real powers to do things. It is also useful to speak of them as enablements or affordances where this conveys the dispositional notion of causal mediation between events. The dispositionalist thus has a commitment to instantiated dispositions that get manifested when a certain set of conditions come into being.

The instantiation of a property in a particular enables the particular to do certain things in certain situations. Thus we have natural necessities without commitment to general laws ranging over classes of events. Instead of general necessities, natural necessity occurs at the level of the particular. It is not in virtue of a general law that sugar dissolves when in liquid, for instance, it is in virtue of a particular state or instantiated property possessed by that sample that it does so. The alternative explanation of disposition manifestations would thus be:

[wv_D] (dispositions + events) ○→ (events describable as disposition manifestations).

Dispositional explanation in this context is not a species of Deductive Nomological explanation because there are no covering laws connecting properties to events. Instead, explanation of events could follow something like the following model:

C_1: *i* was in a situation of kind *S*.
C_2: *i* has the property M that disposes behaviour *R* in a situation of kind *S*.
E: *i* behaved in manner *R*.

The dispositionalist ontology avoids all the problems of laws of nature but does require a convincing account of dispositions as real powers showing that such dispositions can do all that laws were required to do and more besides. We will see that in addition to providing such an ontology another problem the dispositionalist must face is what I call the problem of generality. This problem is how, without general laws, can the dispositionalist explain why generalities in behaviour are true of kinds. All sugar dissolves in liquid and tastes sweet to the tongue. Is it merely a cosmic coincidence that each sample of the kind carries the same set of capacities? Could we have rogue samples that possessed different dispositions?

I will argue that the correct analysis and ontology for dispositions does allow them to usurp laws as the fundamental entities of a convincing metaphysic. I will thus support the dispositionalist against the laws view and give a response to the problem of generality. Before I give the positive case for dispositionalism, however, I will first return to the problems for laws and explain why we are right to overthrow laws of nature as the basic building blocks of explanation.

10.4 *The Problem with Laws*

Two types of general statement may be said to be describable as articulations of laws. The first type is a posteriori identity statements such as 'water is H_2O', 'heat = mean molecular motion', and so on. Such identities are not the kind of law that I am concerned with. I am concerned with another kind of general statements which have the form

$$(L_l) \quad \forall x\,(Fx \;\Box\!\!\rightarrow Gx)$$

such as 'all water boils at 100°C' and 'all ducks have webbed feet'.[3] The idea of such general truths, once we find the fundamental ones, as ultimate and inexplicable explanatory units is a view that is intimately connected with the empiricist tradition of which Hume was a part. The empiricist can make no statement of laws of nature other than that of which is observable, such as the data of experimental evidence. Anything more than this is 'beyond our epistemological grasp'.[4] Because of this the empiricist can only, if they allow any sense to laws at all, identify them with universally quantified conditional statements.

There are a couple of standard objections to such a basic statement of laws. First, this characterization of laws cannot distinguish coincidental from genuinely lawlike regularities and second, virtually no statements of this form are true. Both these problems have resulted in a wealth of literature with the notion of a law being variously defended and then attacked once more.

The first problem arises from the conceivability of true general statements which are clearly not representative of laws. It is no doubt a true regularity that every time I sneeze, someone, somewhere in the world, immediately coughs. There is a constant conjunction of events but we are doubtful that this is because of a law of nature. We may require, for instance, that there be some further connection between those events constantly conjoined but the consistent empiricist, who denies realism about causation, can admit no such further connection.

[3] Another kind of laws are functional laws where one magnitude varies as a function of another. The findings in the case of the kind of laws I do discuss is, I think, generalizable to these cases.

[4] F. Dretske, 'Laws of Nature', *Philosophy of Science*, 44 (1977), 249.

One consistently empiricist response is to attempt to refine the formulation of laws so as to exclude such apparent coincidences. Thus, there may be a demand for a high degree of inductive support for a true general statement to qualify as a law. It may be thought possible, for instance, to test the truth of the general statement in a wide variety of circumstances such that the 'accidental' constant conjunction is shown to fail. Though it may be true that all men in the room wear wristwatches, the attempt to enter the room by a man without a wristwatch would quickly show that the general statement was no more than a coincidence. It is arguable, however, that there is no a priori reason why accidental constant conjunctions may not still pass this, and every other, test. It may be an accidental constant conjunction, for instance, that whenever someone enters the room without a wristwatch, their way is barred.

Another response may be, therefore, that of giving up the strictest empiricist ban on the positing of real causal connections. The affect this has on the understanding of laws of nature is that they do involve natural necessities which coincidental constant conjunctions do not. The attempt to distinguish laws and coincidences in this way may, therefore, overcome the problem of explicating the lawlike connection in a sufficiently strong way, even though it does so at a high price to the empiricist. This step is not, however, sufficient to salvage laws because they still face the second problem of laws as general truths.

Nancy Cartwright has argued that the laws of nature *qua* general statements are, for the most part, idealizations and abstractions that should not be regarded as literally true.[5] Nature rarely witnesses the manifestation in actual events of instances of the fundamental general laws in question. Other events get in the way and the behaviour of other objects intervenes. Even if we do find examples where laws are manifested in an unadulterated way, such as the solar system being a perfect exhibition of the law of gravity in action, the statement of the law claims universal scope and thus is falsified by just one exception.

Because of the possibility of events intervening and stopping the

[5] In particular in her *How the Laws of Physics Lie* (Oxford, 1983), and 'Aristotelian Natures and the Modern Experimental Method', in J. Earman (ed.), *Inference, Explanation and Other Philosophical Frustrations* (Los Angeles, 1993).

manifestation of the law, it may be thought that we can say that general laws are true *ceteris paribus.* Laws of the form 'All *F*s are *G*s', such as 'all ducks have webbed feet' and 'all unsuspended bodies fall to the ground' may not literally be true, in all cases, for some ducks have lost their legs or some may be non-webbed-feet mutants and gliders may ride the thermals. But surely these exceptions are understood through the intervention of other laws and the general statements, though not true in every case, are true *ceteris paribus* or other things being equal. But what does the *ceteris paribus* claim amount to? If the clause in 'all *F*s are *G*s *ceteris paribus*' refers to particular background conditions, that is

(Lc_1) all *F*s are *G*s unless *X*,

for example, 'all objects fall unless filled with helium', then there remains the possibility of a further background condition, which is not specified in Lc_1, rendering it false. It seems that if *ceteris paribus* clauses are to do the job of making the conditional statements true, they must do so by proposing a 'catch-all' stating that all events that will prevent the manifestation of the law are excluded. When this is incorporated into the law, however, it clearly preserves the truth of the law at the expense of making it trivial. Thus 'all *F*s are *G*s *ceteris paribus*' becomes

(Lc_2) all *F*s are *G*s, except when they are not,

for example, 'all objects fall to the ground unless they do not'. If there is nothing between the interpretations Lc_1 and Lc_2—between the false and the trivial—then the *ceteris paribus* strategy is useless.

Cartwright suggests that when citing laws in explanation we are stating what is necessitated in artificial conditions, even though such conditions never (or seldom) occur in reality but only in theoretical models. We are abstracting away from all the messy events that get in the way of a pure instantiation of the law. However, for the purposes of the dispositionalist position it is not necessary to deny that some general statements, perhaps concerning the fundamental elements of science, are true in actuality. There is ample room for debate here but the dispositionalist need not enter it. This is because even if there are some true generalities the dispositionalist can urge, first, that their truth is parasitic upon particular truths and, second, their generality can be accounted for by a position called dispositional essentialism. I will develop these points below (Sects. 10.6 and 10.7, below).

There is a rival interpretation of laws, however, which gets away from the notion of their statements being *general* truths. This, I would accept, is an improved position that is intermediate between the rejected generalities and the acceptable dispositionalist account. This is the Dretske/Armstrong view that dispenses with the empiricist notion of laws being best articulated in general conditional statements connecting types of event and instead construes statements of laws as singular statements of fact that describe relations between properties. Dretske says, for instance, that lawlike statements are actually of the form:

(L_d) F-ness $\rightarrow$ G-ness,

where the connective '$\rightarrow$' can mean different things in the cases of different laws.[6] Thus, these law statements, as true singular statements of fact, will support any universal truths that emerge but are not themselves universal nor implied by any universal truths as the empiricist account requires.

This is an advance. It allows realist causal claims that go beyond the available empirical evidence and it also allows that true general statements are secondary to, or made true by, the actual facts of causation holding between properties. However, the account is one that I would seek to revise and build upon. The Dretske/Armstrong account urges a position of causal connections holding between universals. I suggest, instead, that the dispositions of things can be understood as particulars—states or instantiated properties—such that the real causal connections are connections between particulars rather than universals. The laws view assumption that the fundamental truths are general remains in the Dretske/Armstrong view in virtue of the causal connections holding between universals. The alternative view takes the dispositions, propensities, or capacities of particular things, rather than universals, to be the fundamental units that can be used in explanation of events. Laws, insofar as they are useful as idealizations or abstractions, are built out of these rather than the opposite view which sees dispositions of particulars emerging out of universals together with general laws.

[6] 'Laws of Nature', 253.

10.5 *Prescriptive and Descriptive Laws*

What are formulations of law statements supposed to tell us? They are statements of regularities but what is the status of such regularities? Can they be violated? If not, in what sense are they laws? The most important issue that arises here is whether the laws of nature are a variety of prescriptive or descriptive laws. The most obvious answer to this question is that they are descriptive laws.[7]

The notion of a law of nature as a prescription has obvious connections with the possible existence of a supernatural being that is the lawmaker. For the most part this view is rejected and I shall be making the assumption that we have laws without having a lawmaker. However, the terms in which the debate is phrased can sometimes carry a trace of the prescriptive view of laws. The following considerations suggest that such a prescriptive view should be rejected.

A prescriptive law can be disobeyed; indeed, if it could not be disobeyed there would be no need for it. Thus, there is no prescriptive law against being in two places at the same time or being both over six feet tall and under six feet tall at the same time. Laws in the prescriptive sense are like rules about what should happen in a given situation but they do not state what is logically necessary. Rules were not exactly meant to be broken, but it must be possible that they be broken if they are to have any point. What we take to be the laws of nature are different. Clearly, there is an important sense in which laws of nature cannot be broken but the modality in this case is obviously natural rather than logical. Although it is logically possible that the laws be otherwise, given that they are as they are in fact, then they cannot be disobeyed in the way a rule against dropping litter can be disobeyed. One really has no choice about whether to obey or disobey the laws of nature. For this reason, the more helpful account of laws of nature is that they are descriptive of what actually does happen or, by our theoretical commitments, is expected to happen in the future.

But this is not quite the whole story. The sense in which laws are

[7] There is not unanimous agreement on this. E. J. Lowe argues in favour of laws as prescriptions ('What *is* the "Problem of Induction"?', *Philosophy*, 62 (1987), sect. III). It will be clear from what I say below why that account is rejected.

purely descriptive is yet to be decided. Naïve descriptivism about laws cannot be correct. Laws are something more than mere descriptions of what has happened, or will happen. They are not elliptical for a set of actual past, present, or future events. Standing in the way of such an account would be two problems concerned with how probabilities and unrealized laws would fit with the events of the world's history, as I now detail.

1. *Probabilities.* Many regularities in nature are probabilistic.[8] It is reasonable to assume, however, that probability assignments are objective facts about the world. That a fair coin has an even chance of landing heads or tails, for instance, is reasonably taken to be a fact about the world. Clearly we do take it as such because we can accept the possibility of a fair coin landing on heads more than on tails for any finite set of tosses. This raises a problem for a descriptive account of laws because the probabilistic law, that a fair coin has an equal chance of landing heads or tails, is consistent with any number of heads in a finite sequence of tosses. Clearly, then, the 50/50 probability assignment is not a simple report of the empirical evidence. It seems reasonable to say that a probability of less than 1 and greater than 0 cannot place any restriction on what actually occurs because it is consistent with any distribution of events. Probabilistic laws are thus clearly not descriptions of what has happened in the past or will happen in the future.

2. *Unrealized laws.* The naïve descriptivist account of laws is committed to the view that if two complete world histories are exact in every event, then those two worlds have the same laws. However, given that laws as most frequently formulated are stated in conditionals, then it is possible that there be a law in that world that does not get manifested because the antecedent conditions for that law do not get realized. It may be a law in world $_{w1}$, for instance, that $\forall x\ (F_1 x \circ\!\!\rightarrow F_2 x)$ but for all times in $_{w1}$, $\neg F_1 x$. Given that this is intuitively acceptable, it is possible that in world $_{w2}$, which coincides in its complete world history with $_{w1}$, it is a law that $\forall x\ (F_1 x \circ\!\!\rightarrow \neg F_2 x)$. This difference in law makes no difference to the events in $_{w2}$ because, given that the histories of $_{w1}$ and

[8] We could say that all laws are probabilistic but that some could have a probability of one.

$_{w2}$ coincide, $\neg F_1 x$ is true in $_{w2}$ also. Therefore, two worlds that coincide in all events can differ in their laws.[9]

Naïve descriptivism about laws is not true but clearly the descriptive account of laws seems intuitively preferable to a prescriptive account. How, then, can the descriptive account be defended? It can be defended if general statements are taken as descriptive not of actual events but of some other facts about the world. My candidate item to fill this role is the disposition.

10.6 *An Explanatory Ontology*

How can we have an ontology for dispositions that allows them to occupy an explanatory role superior to that of the covering law model and is without reference to general laws? The short answer to this is that we replace laws of nature with real dispositions as the ultimate, inexplicable units which are the explanations of change. Such explanatory units also show why the argument from potentiality, raised in Sect. 5.7, is no serious threat to the theory of dispositions that has been developed. Disposition instantiations are identical to certain categorical instantiations even though it is possible to know one particular categorical property of an object without thereby knowing what dispositions that object has. Until now, the explanation of this possibility was in terms of dispositions, for any particular categorical state, being dependent upon contingent laws of nature. This is rephrased in a way which does not depend upon laws, though it does depend upon contingency (Sect. 10.8). Properties or states are now understood to contain real potentialities.[10]

This needs further development however, for the majority of

[9] In 'A Subjectivist's Guide to Objective Chance', postscripts B and C in *Philosophical Papers*, ii, David Lewis makes a stout defence of Humeanism against such problems. He ultimately finds acceptable, while the present author does not, first, that chances must supervene on the history of particular occurrences and, second, that two worlds that coincide in their occurrences have the same laws (laws being best-possible systematizations that have only true consequences). The detail of Lewis's argument is tackled head-on by Robert Black, 'Chance, Supervenience and the Principal Principle' (forthcoming).

[10] Thus their behaviour can be extended to hypothetical cases, contrary to an objection made by M. Tooley, *Causation* (Oxford, 1987), 70. Dispositional claims are not based upon the actual behaviour of the things which just happen to exist, as shown in the previous section.

cases of dispositions, what may be called 'ordinary' dispositions, are not ultimate and inexplicable; indeed in many cases we are pretty sure that we do have an explanation of the mechanism involved in the disposition manifestation. We know, for instance, why a clock succeeds in telling the right time, why a billiard ball is disposed to roll in a straight line when struck, and even why sugar dissolves in liquid. The explanations we can give for these dispositions may be called structural or, in the terms stated in the last chapter, homunctional. This is an explanation of the actual or possible behaviour of the whole in terms of the actual and possible behaviour of its parts. The same type of explanation could be repeated for the lower levels of the components but eventually, if the strategy is repeated, a level will be reached where no further explanation is possible because there is no further structural or homunctional explanation available for that capacity. Such a capacity is what I have called an ungrounded disposition and it is such ungrounded dispositions which I suggest are fit to replace laws of nature as the basic, unexplained units of explanation. Five things need to be said about the role of these dispositions to make clear the plausibility of the dispositionalist ontology.

First, most dispositions we encounter are of the ordinary variety. Here there is no need to appeal to ultimate powers because explanations of the behaviour enabled by such dispositions can be explained by reference to the components that constitute that disposition. The cases where we appeal to ultimate dispositions are therefore limited: there are not as many ultimate dispositions as there are dispositions. A point of parsimony is involved here. Just as we may regard it as good methodology to reduce laws of nature to those that are fundamental and capable of explaining other laws, which are less fundamental, so we may regard it as good methodology to posit no ultimate disposition unless we have to, that is, unless we have no explanation or idea of what could possibly be an explanation. Hence the permission of unexplained ultimate dispositions is no sanction for unnecessarily populating the world with the occult *Qualities* that Boyle was at pains to eliminate. The dispositions that we allow to be ultimate are therefore a limited class, examples of which would be the dispositions of subatomic particles which apparently have no internal constitution that could explain their behaviour. There are, for instance, two types of subatomic particles that are called *muons* and *tauons*. The only difference between a muon and a tauon is in their behaviour. Paul Davies explains that it is in virtue of this, and only this, that these particles are classified as they are: 'The vital statistics of subatomic particles are mass, electric charge, and spin.'[11] No other difference between the particles exists in the theory because such dispositional behaviour is arguably the only property each type of particle possesses. Muons and tauons are both leptons and leptons are understood as lacking in any internal structure that explains their possession of these dispositions. Hadrons do have internal structures that cause and explain their differences in behaviour. They are composed of combinations of quarks. Quarks were initially thought to come in three varieties: up quarks, down quarks, and strange quarks. These are distinguished by their differences in behaviour; for example, one difference they can have is a difference in charge. To this list have been added three other varieties: charm, bottom, and top. Such quarks are currently thought to be the basic building blocks of all hadrons and together with leptons these are thought to be the simple atoms, which lack internal structure, of which all matter is constituted. The properties these things bear are the best examples of ungrounded dispositions that can currently be offered.

The second point to be made about the dispositionalist ontology is that laws, *qua* true generalities, if they exist at all, are ontologically parasitic upon the capacities of particulars, rather than the other way round. This reverses the way many Humean detractors of the dispositional have understood the direction of dependence. The dispositionalist can accept that some true generalities exist, or generalities with *ceteris paribus* qualifications, but give a fundamentally anti-Humean account of these generalities. They are not the fundamental, inexplicable truths the Humean takes them to be, rather they are generalizations over classes of dispositions. Particulars do not do things because there are laws; there are laws because there are things that particulars (can) do. Put another way, the Humean understands the dispositions of things to be supervenient upon laws; the dispositionalist takes laws to be supervenient upon dispositions.

The direction of dependence between laws and dispositions is thus turned about and this leads on to the third point. Such laws,

[11] *Superforce*, 81.

qua true generalities, are descriptive, in the sense discussed above; but what it is that they are best understood as descriptive of is not actual or possible events but the capacities of things. Thus the truth of the claim that a fair coin has a 50/50 chance of landing a head or a tail is true not in virtue of the result of any toss or sequence of tosses, but in virtue of the 50/50 propensity the coin has to land head or tail. Clearly this is not an ultimate disposition however, an explanation in terms of other properties of the coin—shape and weight distribution—is readily available. The probabilistic law can thus be explained by it being descriptive of an actual disposition and the account also explains the possibility of unrealized laws. The disposition ascription is true of the coin even if it is never tested and thus there can be two true but different disposition ascriptions in two worlds that coincide in all their events. Given the realist understanding of such dispositions, this would be something more than a counterfactual difference.

The fourth point I make concerns the warrant we have for claims that a particular disposition is ultimate rather than ordinary. The history of science shows that mistakes are possible in this respect. Atoms may have been thought to be the smallest possible units, lacking any internal structure that could explain their behaviour. This is now rejected but they have been replaced by further units—leptons and quarks—that are instead alleged by many physicists to be the smallest units, lacking any structural components.[12] We ought, therefore, to proceed with caution. We cannot be sure, for any putative ungrounded disposition, whether it is genuinely ungrounded or actually just an ordinary disposition and only thought to be ultimate because of the incomplete state of our physics. I will call such dispositions 'epistemically ungrounded' where a structural explanation exists of which we have yet to attain knowledge. If a disposition is epistemically ungrounded, then it is still, in practice, the thing to which finally we appeal in explanation and which is one of the basic building blocks from which we construct law statements. Our explanation ends here, even though it is possible that at a later stage we advance beyond this boundary to a deeper level of explanation. When such a

[12] This view is currently more popular than the waning superstring theory but what I say of leptons and quarks is equally applicable to the units appealed to in this alternative view.

boundary is passed then we may say, retrospectively, that the epistemically ungrounded disposition played a mere heuristic role: it offered us a possibility of research. A new boundary will be set, however. Some new boundary of ungrounded explanation will be laid down, even though this level, in turn, may be found to be only epistemically ungrounded.

Could all dispositions be merely epistemically ungrounded and hence 'ordinary' rather then genuinely ungrounded? It seems that there is no a priori reason why this cannot be the case but with the acceptance of such a view comes the acceptance of an infinite number of ever smaller and more basic levels of entity. This may be regarded as counter-intuitive but it would not be the first time that physical theory and common sense were at odds. The alternative may equally be regarded as counter-intuitive, though. It may be wondered how genuinely ungrounded dispositions could possibly exist but can their existence be ruled out a priori? Apparently not. Again, common sense and physical theory may prove to be out of step.

However, the fifth point I wish to make is that the nature of explanation is such that ungrounded dispositions will always have to be *posited* in order to avoid a regress of explanation and, further, that we have every reason, as part of the atomistic strategy, to assume that there are genuine ungrounded dispositions. The sort of atomistic explanations that we have been striving for in science must always be based on the assumption of units which are themselves unexplained. These, in turn, may be explained in due course but unless we accept a unit as itself inexplicable, we will never complete, even temporarily, a process of explanation. It is just this role, I have argued, that is filled in our explanation of the physical world by ungrounded dispositions.

10.7 *Generality*

One benefit that it seems laws can provide us, that capacities cannot, is generality. Perhaps this feature is so important to us that we will still attempt to defend laws in the face of the arguments thus far. Laws hold over whole classes of particular and bind them to certain types of behaviour in certain types of situation. They not only account for regularities in the behaviour of particulars over time but also regularities in the behaviour of *types* of particular at a time. We could call these diachronic and synchronic regularities of nature respectively and express them in the questions: why, without laws, should objects have the same dispositions tomorrow that they have today and why, without laws, should the same properties in different objects support the same dispositions?[13] Although one sample of sugar may be disposed to dissolve when placed in liquid, for instance, another sample, of otherwise indiscernibly similar sugar, would be disposed to ignite when in the same conditions. Capacities, individual to each particular thing, bring no guarantee of consistent behaviour in kinds, at a time or over time, and, the Humean argues, this is one point on which laws are enormously advantaged.

The dispositionalist can answer these charges. That generalities over time and at a time can hold in a world of particular capacities will be explained by a two-stage strategy. First, regularities among large-scale classes of particular will be shown to stem from the regularities among the capacities of their component sub-capacities. Second, the regularities of sub-capacities will be shown to rest finally on the ultimate, ungrounded dispositions for which we can adopt a position of dispositional essentialism. Dispositional essentialism, it will be explained, ensures regularities but not in the way the laws view would have it.

First, let us consider ordinary macroscopic dispositions of medium-sized objects, such as the disposition of a grandfather clock to tell the right time if its pendulum is swinging freely. If we were to attempt to explain this behaviour by appeal to ultimate dispositions possessed by each clock, then it seems a logical possibility that such clocks differ radically in behaviour. But the behaviour of clocks would not be explained by appeal to ultimate dispositions because structural explanations are readily available for a clock's ability to tell the right time. Further, if one clock was running fast, a clock-maker could diagnose the problem and perhaps isolate a faulty component or set of components. The principle of homunctionalism is invoked. The disposition of the whole is regulated by the dispositions of the component parts of which it is made. These make causal contributions to the whole which add up to one big

[13] These questions are raised against a dispositionalist strategy by N. Everitt, 'Strawson on Laws and Regularities', *Analysis*, 50 (1991).

effect: the regular movement of the hands. That a number of clocks are alike in their behaviour is because they are alike in having components, not necessarily of the same categorical nature, which add up to produce the same kind of effect. Thus, given the dispositions of the parts, the whole could not behave in a way other than it does.

The homunctionalist strategy clearly has more appeal than the laws one here. In answer to the question of why a particular behaves in a certain way the supporter of laws cites a general law governing all particulars of that kind. But it is not such a truth, if it is a truth, that exerts an influence on an object; it is the particular instantiations of properties—facts about the object concerned—that do the causal work. These are particular property instantiations possessed by particular objects which make those objects have particular dispositions.

Higher-level dispositions can be explained by appeal to lower-level dispositions. The same strategy can be used until we reach a level where we have dispositions with no components, hence no components in common with other bearers of the same disposition. These are the genuinely ungrounded dispositions that have no internal structure or the epistemically ungrounded dispositions for which the internal structure is unknown. If such capacities are to be taken as 'brute' facts that cannot be explained homunctionally, then why, say, could two electrons not behave in completely different ways? Dispositional essentialism provides the answer. Basically, an electron is an electron solely in virtue of its dispositions to behave; hence anything that was not disposed to behave in this kind of way would not be an electron but some other kind of subatomic entity.[14] Note that this is not the same as saying that classification into kinds is based on behaviour solely. There may be a probabilistic disposition involved where two different forms of behaviour can be manifested randomly at different times by objects of the same kind. In this case, the analysis of probabilistic dispositions is pertinent and two such objects can be classed among the same kind on the basis of having the same probabilistic dispositions to behave though not the same actual behaviour.

[14] In my 'Ellis and Lierse on Dispositional Essentialism', I endorsed dispositional essentialism for certain basic kinds though I took to task Ellis and Lierse, 'Dispositional Essentialism', for their ontology which they deemed necessary for the position.

There is, thus, a theoretical commitment involved here which concerns the basis upon which certain kinds are identified. The justification for this basis of classification in the case of subatomic particles is that the dispositions to behave are the only features of those things on which a classification could be based. Thus regularities among classes do exist but not because there is a law stating that all of kind *K* have disposition *D*. Rather, regularities exist because we make our classification of kinds, at this level, on the basis of their dispositions, not on any pre-established division into kinds. The dispositions are actually the bases for classification into kinds and this is why all electrons behave in the same way.

Regularity takes its toehold at this level. Out of such microscopic regular dispositions spring regularities in larger-scale dispositions. Macroscopic-sized objects exhibit the same behaviour when they have the same dispositional components. When they have different dispositional components, similarity in behaviour is not expected and differences in behaviour are accepted.

Erik Anderson accuses me of failing to prove the 'basic laws' view in my reply to Ellis and Lierse's discussion of dispositional essentialism.[15] However, such a proof of the basic laws view was not my purpose. Rather, the charge Anderson brings against me is the same charge that I brought against Ellis and Lierse. They had argued that dispositional essentialism entailed a definite answer to this debate: that dispositions were ultimate, rather than laws. This involved them making a number of controversial claims, such as that the laws of nature were logically necessary instead of contingent as the Humean tradition takes them to be. Ellis and Lierse did not, however, have anything by way of convincing positive argument for their view: they had set out a view without adequately justifying it. I therefore stand by my original charge that the realist theory of dispositions, which Ellis and Lierse take to be necessary for dispositional essentialism, is neither proved by their arguments nor required by the position and nothing said by Anderson undermines this accusation. My aim was to show how dispositional essentialism could be true even if it were laws rather than dispositions that were in the relevant sense ultimate or basic. So Ellis and

[15] E. Anderson, 'Dispositional Essentialism: Alive and Well', *Philosophical Papers*, 26 (1997), 195–201.

Lierse have not disproved the basic laws view but this does not mean that the basic laws view is *proved* or that I was offering such a proof, even though in places I pushed that view for dialectical purposes. Further, although basic dispositions as truthmakers of laws are not necessary for the claim of dispositional essentialism about certain fundamental kinds in science, as Ellis and Lierse suggested, what I say above shows that dispositional essentialism may be required for the credibility of the basic dispositions view because it suggests a solution to the generality problem that exists for the basic dispositions view. Hence, basic dispositions are not required by dispositional essentialism but dispositional essentialism may be required for basic dispositions.

10.8 *Contingency*

There is a strong and sensible intuition that laws of nature are logically contingent. Let us call this the contingency thesis (CT).

CT: the laws of nature are logically contingent.

I have implicitly and explicitly endorsed the contingency thesis at various places earlier. Although dispositions bring natural necessity into the world there still exists the kind of logical contingency that is required. The replacement of laws with real dispositions is no threat to the view that the way nature behaves could have been otherwise but it does mean that CT will have to be revised accordingly. A modified version of the contingency thesis can be given that is framed in terms of the contingency of a particular's capacities rather than the contingency of laws. There exist relevant conceptual necessities of dispositional essentialism, such as that described above where an electron must, conceptually, have certain dispositions if it is to be an electron. Such conceptual necessities do not threaten natural contingency in the sense that the only logical necessities are conceptual. Hence it is logically contingent that a particular entity is a particular with the capacities it actually has. That particular could have had different dispositions to the ones it actually has, even though this would entail that the particular belonged to a different kind in virtue of that fact. Such contingency does not threaten the identities for such entities across possible worlds however, as I have shown previously, for our disposition ascriptions are actual-world-relative.[16]

Ellis and Lierse argued that a position of dispositional essentialism required the logical necessity of physical laws. This contentious move is, I think, unnecessary. While it is certainly true that an electron would not be an electron if its behaviour were different from the behaviour it has in the actual world, this necessity is purely conceptual. That it is in virtue of behaviour B that a particular x is classified within a kind K does not entail that x necessarily has behaviour B. There is no contradiction in saying that x could have had behaviour B', other than it actually has, and thus could have belonged to a different kind K'.

The Ellis/Lierse view involves a misunderstanding of the scope of the logical necessity involved. From the conceptual necessity

$$\Box \forall x \,(x \text{ is an electron} \leftrightarrow x \text{ has behaviour } B)$$

it does not follow, for any particular x, that x has behaviour B necessarily. That a particular possesses any disposition is logically contingent even though some particulars, such as electrons, would not have been classed as such if they had different behaviour. To deny this would be to claim that an electron's behaviour is dictated by logic and, presumably, physics is a trivially analytic human folly.[17]

This means that a revised contingency thesis (RCT), expressed in terms of dispositions instead of laws, can be supported:

RCT: which dispositions are possessed by particulars is logically contingent (though for some kinds K_1, K_2, . . . K_n, if x has appropriate dispositions, necessarily x is a member of those kinds).

Therefore, the way in which the dispositional–categorical distinction was made in Chapter 4 can be preserved in essence though altered in detail. A disposition ascription supports a conditional as a matter of conceptual necessity. A categorical ascription does not but this is explained not in terms of the contingency of laws of nature but in terms of the contingent relation between a categorical property and what is enabled by the possession of such a property.

[16] 'Ellis and Lierse on Dispositional Essentialism', 607–8.

[17] Hence my dispositional essentialism differs from the otherwise commendable version presented by Ellis and Lierse.

The considerations advanced in the present chapter weigh in favour of replacing general laws which bind classes to certain modes of behaviour with real dispositions that bind individuals. Throughout I have been trying to show how such a world view is not only one that can reasonably be supported but also that it is one which is in tune with a commonsense way of thinking about the world. It seems that there is not enough that general laws give us that cannot be provided at least equally well by individual dispositions. If this discussion has been on track, then the significance of the conceptual and ontological questions surrounding dispositions, that have been discussed in this study, is promoted to paramount importance.

BIBLIOGRAPHY

Alston, W. P. (1971), 'Dispositions, Occurrences, and Ontology', in R. Tuomela (ed.), *Dispositions*, Dordrecht, Reidel (1978), 359–88.

Anderson, E. (1997), 'Dispositional Essentialism: Alive and Well', *Philosophical Papers*, 26: 195–201.

Aristotle, *The Metaphysics*, in *The Works of Aristotle translated into English*, viii, trans. and ed. by W. D. Ross, Oxford, Clarendon Press, 1908.

Armstrong, D. M. (1968), *A Materialist Theory of the Mind*, London, Routledge.

—— (1969), 'Dispositions are Causes', *Analysis* 30: 23–6.

—— (1970), 'The Nature of Mind', in C. V. Borst (ed.), *The Mind/Brain Identity Theory*, London, Macmillan; repr. in N. Block (ed.), *Readings in Philosophy of Psychology*, i, London, Methuen, 191–9.

—— (1973), *Belief, Truth, and Knowledge*, Cambridge, Cambridge University Press.

—— (1978), *A Theory of Universals*, Cambridge, Cambridge University Press.

—— (1983), *What is a Law of Nature?*, Cambridge, Cambridge University Press.

—— (1989), 'C. B. Martin, Counterfactuals, Causality and Conditionals', in J. Heil (ed.), *Cause, Mind and Reality: Essays Honoring C. B. Martin*, Dordrecht, Kluwer, 7–15.

—— Place, U. T., and Martin, C. B. (1996), *Dispositions: A Debate*, London, Routledge.

Attneave, F. (1960), 'In Defense of Homunculi', in W. Rosenblith (ed.), *Sensory Communication*, Cambridge, Mass., MIT Press.

Ayers, M. R. (1975), 'The Ideas of Power and Substance in Locke's Philosophy', in I. C. Tipton (ed.), *Locke on Human Understanding*, Oxford, Oxford University Press, 77–104.

Bacon, J. (1996), *Universals and Property Instances*, Oxford, Blackwell.

Berg, J. (1955), 'On Defining Dispositional Predicates', *Analysis*, 15: 85–9.

Bergmann, G. (1955), 'Dispositional Properties and Dispositions', *Philosophical Studies*, 6: 77–80.

Black, R. (forthcoming), 'Chance, Supervenience and the Principal Principle', *British Journal for the Philosophy of Science*.

Blackburn, S. (1990), 'Filling in Space', *Analysis*, 50: 62–5.

Block, N. (1980), 'What is Functionalism?', in N. Block (ed.), *Readings in Philosophy of Psychology*, i, London, Methuen, 171–84.

Block, N. (1994), 'Functionalism (2)', in S. Guttenplan (ed.), *A Companion to the Philosophy of Mind*, Oxford, Blackwell.

Bochenski, I. M. (1961), *A History of Formal Logic*, trans. by I. Thomas, Notre Dame, University of Notre Dame Press.

Boyle, R. (1666), *The Origin and Forms and Qualities*, in M. A. Stewart (ed.), *Selected Philosophical Papers of Robert Boyle*, Manchester, Manchester University Press, 1979, 1–96.

—— (1674), 'About the Excellency and Grounds of the Mechanical Hypothesis', in M. A. Stewart (ed.), *Selected Philosophical Papers of Robert Boyle*, Manchester, Manchester University Press, 1979, 138–54.

Bricke, J. (1975), 'Hume's Theory of Dispositional Properties', *American Philosophical Quarterly*, 10: 15–23.

Broad, C. D. (1925), *The Mind and its Place in Nature*, London, Harcourt Brace.

Butler, D. (1988), 'Character Traits in Explanation', *Philosophy and Phenomenological Research*, 49: 215–38.

Campbell, K. (1990), *Abstract Particulars*, Oxford, Blackwell.

Carnap, R. (1932), 'Die Physikalische Sprache als Universalsprache der Wissenschaft', trans. by M. Black as *The Unity of Science*, London, Kegan Paul (1934); also in O. Hanfling (ed.), *Essential Readings in Logical Positivism*, Oxford, Blackwell (1981), 150–60.

—— (1936), 'Testability and Meaning' (part I), *Philosophy of Science*, 3: 420–71.

—— (1937), 'Testability and Meaning' (part II), *Philosophy of Science*, 4: 1–40.

—— (1956), 'The Methodological Character of Theoretical Concepts', *Minnesota Studies in the Philosophy of Science*, 1: 38–76.

Cartwright, N. (1983), *How the Laws of Physics Lie*, Oxford, Clarendon Press.

—— (1993), 'Aristotelian Natures and the Modern Experimental Method', in J. Earman (ed.), *Inference, Explanation and Other Philosophical Frustrations*, Los Angeles, University of California Press, 44–71.

Champlin, T. S. (1991), 'Tendencies', *Proceedings of the Aristotelian Society*, 1990–1: 119–33.

Churchland, P. M. (1981), 'Eliminative Materialism and the Propositional Attitudes', *Journal of Philosophy*, 78: 67–90.

Coder, D. (1969), 'Some Misconceptions about Dispositions', *Analysis*, 29: 200–2.

Cummins, R. (1974), 'Dispositions, States and Causes', *Analysis*, 34: 194–204.

—— (1975), 'Functional Analysis', *Journal of Philosophy*, 72: 741–65.

D'Alessio, J. C. (1967), 'Dispositions, Reduction Sentences and Causal Conditionals', *Critica Revista Hispano Americana de Filosofia*, 14: 65–76.

Dalrymple, H. B. (1975), 'Dispositional and Causal Explanation', *South-West Journal of Philosophy*, 6: 115–21.

Daly, C. (1994), 'Tropes', *Proceedings of the Aristotelian Society*, 94: 253–61.

Davidson, D. (1963), 'Actions, Reasons and Causes', in *Essays on Actions and Events*, Oxford, Oxford University Press (1980), 3–19.

Davies, P. (1995), *Superforce*, revd. edn., London, Penguin.

Dennett, D. C. (1971), 'Intentional Systems', *Journal of Philosophy*, 68: 87–106.

—— (1975), 'Why The Law of Effect Will Not Go Away', in W. G. Lycan (ed.), *Mind and Cognition*, Oxford, Blackwell (1990), 63–77.

—— (1991), *Consciousness Explained*, London, Allen Lane, 1992.

Descartes, R. (1641), *Meditations on First Philosophy*, in J. Cottingham, R. Stoothoff, and D. Murdoch (eds.), *The Philosophical Writings of Descartes*, ii, Cambridge, Cambridge University Press (1985), 1–62.

Dicker, G. (1977), 'Primary and Secondary Qualities: A Proposed Modification of the Lockean Account', *Southern Journal of Philosophy*, 15: 457–71.

Dretske, F. (1977), 'Laws of Nature', *Philosophy of Science*, 44: 248–68.

Dummett, M. (1977), *Elements of Intuitionism*, Oxford, Clarendon Press.

—— (1978), *Truth and Other Enigmas*, London, Duckworth.

—— (1982), 'Realism', *Synthese*, 52: 55–112.

Ellis, B., and Lierse, C. (1994), 'Dispositional Essentialism', *Australasian Journal of Philosophy*, 72: 27–45.

Everitt, N. (1991), 'Strawson on Laws and Regularities', *Analysis*, 50: 206–8.

Fales, E. (1990), *Causation and Universals*, London, Routledge.

Fetzer, J. H. (1977), 'A World of Dispositions', *Synthese*, 34: 397–421; also in R. Tuomela (ed.), *Dispositions*, Dordrecht, Reidel (1978), 163–87.

Feyerabend, P. (1963), 'Mental Events and the Brain', *Journal of Philosophy*, 60: 160–6.

Franklin, J. (1988), 'Are Dispositions Reducible to Categorical Properties?', *Philosophical Quarterly*, 38: 62–4.

Frege, G. (1879), *Begriffsschrift*, trans. by T. W. Bynam as *Conceptual Notation*, Oxford, Clarendon Press (1972).

—— (1892), 'Über Sinn und Bedeutung', *Zeitschrift für Philosophie und Philosophische Kritik*, 100: 25–50; trans. as 'On Sense and Meaning' in P. Geach and M. Black (eds.), *Translations from the Philosophical Writings of Gottlob Frege*, 3rd edn., 1980, Totowa, New Jersey, Barnes & Noble, 56–78.

Geach, P. (1957), *Mental Acts*, London, Routledge & Kegan Paul.

Goodman, N. (1955), *Fact, Fiction, and Forecast*, Indianapolis, Bobbs-Merrill Co.; 2nd edn (1965).

Hanfling, O. (ed.) (1981), *Essential Readings in Logical Positivism*, Oxford, Blackwell.

Harré, R. (1970), 'Powers', *British Journal for the Philosophy of Science*, 21: 81–101.

—— and Madden, E. H. (1975), *Causal Powers*, Oxford, Oxford University Press.

Harvey, J. (1992), 'Challenging the Obvious: The Logic of Colour Concepts', *Philosophia*, 21: 277–94.

Hempel, C. G. (1962), 'Explanation in Science and in History', in R. G. Colodny (ed.), *Frontiers of Science and Philosophy*, London, Allen & Unwin, 7–33.

—— (1978), 'Dispositional Explanation', in R. Tuomela (ed.), *Dispositions*, Dordrecht, Reidel (1978), 137–46.

Hume, D. (1739–40), *A Treatise of Human Nature*, L. A. Selby-Bigge edn., Oxford, Clarendon Press (1888).

Hutchison, K. (1991), 'Dormitive Virtues, Scholastic Qualities, and the New Philosophies', *History of Science*, 29: 245–78.

Jackson, F., and Pargetter R. (1987), 'An Objectivist's Guide to Subjectivism about Colour', *Review of International Philosophy*, 41: 127–41.

Joske, W. D. (1967), *Material Objects*, London, Macmillan.

Kripke, S. (1971), 'Identity and Necessity', in M. Munitz (ed.), *Identity and Individuation*, New York, New York University Press (1971), 160–4. Excerpt in N. Block (ed.), *Readings in Philosophy of Psychology*, i, London, Methuen, 144–7.

Lange, M. (1994), 'Dispositions and Scientific Explanation', *Pacific Philosophical Quarterly*, 75: 108–32.

Levi, I., and Morgenbesser, S. (1964), 'Belief and Disposition', *American Philosophical Quarterly*, 1: 221–32; also in R. Tuomela (ed.), *Dispositions*, Dordrecht, Reidel (1978), pp. 389–410.

Lewis, D. (1966), 'An Argument for the Identity Theory', *Journal of Philosophy*, 63: 17–25; also in *Philosophical Papers*, i, Oxford, Oxford University Press (1983), 99–107.

—— (1972) 'Psychophysical and Theoretical Identifications', *Australasian Journal of Philosophy*, 50: 249–58.

—— (1973), *Counterfactuals*, Oxford, Blackwell.

—— (1980), 'Mad Pain and Martian Pain', in N. Block (ed.), *Readings in Philosophy of Psychology*, i, London, Methuen (1980), 216–22.

—— (1980*a*), 'A Subjectivist's Guide to Objective Chance' in *Philosophical Papers*, ii, Oxford, Oxford University Press, 83–132.

—— (1986), 'Causal Explanation', in *Philosophical Papers*, ii, Oxford, Oxford University Press, 214–40.

—— (1997), 'Finkish Dispositions', *Philosophical Quarterly*, 47: 143–58.

Locke, J. (1690), *An Essay Concerning Human Understanding*, Oxford, Clarendon Press (1975).

Lowe, E. J. (1980), 'Sortal Terms and Natural Laws', *American Philosophical Quarterly*, 17: 253–60.

—— (1982), 'Laws, Dispositions and Sortal Logic', *American Philosophical Quarterly*, 19: 41–50.

—— (1987), 'What *is* the "Problem of Induction"?', *Philosophy*, 62: 325–40.

—— (1995), 'The Truth about Counterfactuals', *Philosophical Quarterly*, 45: 41–59.

Lycan, W. G. (1987), 'The Continuity of Levels of Nature', in W. G. Lycan (ed.), *Mind and Cognition*, Oxford, Blackwell (1990), 77–96.

McGinn, C. (1975), 'A Note on the Essence of Natural Kinds', *Analysis*, 35: 177–83.

Mackie, J. L. (1973), *Truth, Probability and Paradox*, Oxford, Oxford University Press.

—— (1977), 'Dispositions, Grounds and Causes', *Synthese*, 34: 361–9; also in R. Tuomela (ed.), *Dispositions*, Dordrecht, Reidel (1978), 99–107.

McLaughlin, B. P. (1995), 'Disposition', in J. Kim and E. Sosa (eds.), *A Companion to Metaphysics*, Oxford, Blackwell (1995), 121–4.

McMullin, E. (1978), 'Structural Explanation', *American Philosophical Quarterly*, 15: 139–47.

Martin, C. B. (1984), 'Anti-Realism and the World's Undoing', *Pacific Philosophical Quarterly*, 65: 3–20.

—— (1993), 'The Need for Ontology: Some Choices', *Philosophy*, 68: 505–22.

—— (1993), 'Power for Realists', in J. Heil (ed.), *Cause, Mind and Reality: Essays Honoring C. B. Martin*, Dordrecht, Kluwer (1989), 175–86.

—— (1994), 'Dispositions and Conditionals', *Philosophical Quarterly*, 44: 1–8.

Mellor, D. H. (1974), 'In Defense of Dispositions', *Philosophical Review*, 83: 157–81.

—— (1982), 'Counting Corners Correctly', *Analysis*, 42: 96–7.

Molière, J.-B. (1682), *La Malade Imaginaire*, in *The Plays of Molière*, viii, trans. by A. R. Waller, Edinburgh (1926).

Montuschi, E. (1991), 'From Effects to Causes: The Role of "Structure" in Scientific Explanation', *Conceptus*, 25: 21–36.

Mumford, S. D. (1994), 'Dispositions, Concepts, and Ontologies', in H. Wallis (ed.), *Language and Related Matters*, Dept. of Philosophy, University of Leeds, 43–5.

—— (1994), 'Dispositions, Supervenience, and Reduction', *Philosophical Quarterly*, 44: 419–38.

—— (1995), 'Dispositions, Bases, Overdetermination and Identities', *Ratio*, N.S. 8: 42–62.

Mumford, S.D. (1995), 'Ellis and Lierse on Dispositional Essentialism', *Australasian Journal of Philosophy*, 73: 606–12.

—— (1996), 'Conditionals, Functional Essences and Martin on Dispositions', *Philosophical Quarterly*, 46: 86–92.

Nagel, E. (1961), *The Structure of Science*, London, Routledge & Kegan Paul.

O'Shaughnessy, B. (1970), 'The Powerlessness of Dispositions', *Analysis*, 31: 1–15.

Pap, A. (1958), 'Disposition Concepts and Extensional Logic', *Minnesota Studies in the Philosophy of Science*, 2: 196–224; also in R. Tuomela (ed.), *Dispositions*, Dordrecht, Reidel (1978), pp. 27–54.

—— (1963), 'Reduction Sentences and Disposition Concepts', in P. A. Schillp (ed.), *The Philosophy of Rudolf Carnap*, La Salle, Ill., Open Court (1963), 559–97.

Peacocke, C. (1978), *Holistic Explanation*, Oxford, Clarendon Press.

Peirce, C. S. (1885), 'On the Algebra of Logic: A Contribution to the Philosophy of Notation', *American Journal of Mathematics*, 7: 180–202; also in C. Hartshorne and P. Weiss (eds.), *Collected Papers of Charles Sanders Peirce*, iii, Cambridge, Mass., Harvard University Press (1933), 210–49.

Place, U. T. (1996), 'Intentionality as the Mark of the Dispositional', *Dialectica*, 50: 91–120.

Plato, *Sophist*. Translation by F. M. Cornford in E. Hamilton and H. Cairns (eds.), *The Collected Dialogues of Plato*, Princeton, NJ, Princeton University Press (1961), 957–1017.

—— *Laches* in T. J. Saunders (ed.), *Early Socratic Dialogues*, London, Penguin (1981), 83–115.

Popper, K. R. (1957), 'The Propensity Interpretation of the Calculus of Probability, and the Quantum Theory', in S. Körner (ed.), *Observation and Interpretation*, London, Butterworth, 65–70.

—— (1959), *The Logic of Scientific Discovery*, revd. imp. 1980, London, Hutchinson.

Prior, E. W. (1982), 'The Dispositional/Categorical Distinction', *Analysis*, 42: 93–6.

—— (1985), *Dispositions*, Aberdeen, Aberdeen University Press.

—— Pargetter, R., and Jackson, F. (1982), 'Three Theses about Dispositions', *American Philosophical Quarterly*, 19: 251–7.

Putnam, H. (1960), 'Minds and Machines', in *Mind, Language and Reality: Philosophical Papers*, ii, Cambridge, Cambridge University Press (1975), 362–85.

—— (1967), 'The Mental Life of Some Machines', in *Mind, Language and Reality: Philosophical Papers*, ii, Cambridge, Cambridge University Press (1975), 408–28.

—— (1967), 'The Nature of Mental States', repr. in N. Block (ed.), *Readings in Philosophy of Psychology*, i, London, Methuen, 223–31.

Quine, W. V. O. (1953), *From a Logical Point of View*, Cambridge, Mass., Harvard University Press.

—— (1960), *Word and Object*, Cambridge, Mass., MIT Press.

—— (1966), *The Ways of Paradox and Other Essays*, revd. edn., Cambridge, Mass., Harvard University Press.

—— (1969), *Ontological Relativity and Other Essays*, New York, Colombia University Press.

—— (1970), *Philosophy of Logic*, Englewood Cliffs, NJ, Prentice-Hall.

—— (1974), *Roots of Reference*, La Salle, Ill., Open Court.

—— (1979), 'Facts of the Matter', in R. W. Shanan and C. V. Swoyer (eds.), *Essays on the Philosophy of W. V. O. Quine*, Hassocks, Harvester (1979), 155–69.

Robinson, H. (1982), *Matter and Sense*, Cambridge, Cambridge University Press.

Russell, B. A. W. (1911/12), 'On the Relation of Universals and Particulars', *Proceedings of the Aristotelian Society*, 12: 1–24.

Russell, L. J. (1934), 'Communication and Verification', *Proceedings of the Aristotelian Society*, Suppl. 13: 174–93.

Ryle, G. (1949), *The Concept of Mind*, London, Hutchinson.

Shoemaker, S. (1980), 'Causality and Properties', in P. van Inwagen (ed.), *Time and Cause*, Dordrecht, Reidel (1980), 109–35; also in *Identity, Cause, and Mind*, Cambridge, Cambridge University Press (1982), 206–33.

Shope, R. K. (1978), 'The Conditional Fallacy in Contemporary Philosophy', *Journal of Philosophy*, 75: 397–413.

Smith, P. (1992), 'Modest Reductions and the Unity of Science', in D. Charles and K. Lennon (eds.), *Reduction, Explanation, and Realism*, Oxford, Clarendon Press (1992), 19–43.

Sober, E. (1982), 'Dispositions and Subjunctive Conditionals, or, Dormative Virtues are no Laughing Matter', *Philosophical Review*, 91: 591–6.

—— (1985), 'Putting the Function Back into Functionalism', in W. G. Lycan (ed.), *Mind and Cognition*, Oxford, Blackwell (1990), 97–106.

Squires, R. (1969), 'Are Dispositions Causes?', *Analysis*, 29: 45–7.

—— (1970), 'Are Dispositions Lost Causes?', *Analysis*, 31: 15–18.

Stalnaker, R. (1968), 'A Theory of Conditionals', repr. in F. Jackson (ed.), *Conditionals*, Oxford, Oxford University Press (1991), 28–45.

Stevenson, L. (1969), 'Are Dispositions Causes?', *Analysis*, 29: 197–9.

Storer, T. (1951), 'On Defining "Soluble"', *Analysis*, 11: 134–7.

Strawson, P. F. (1959), *Individuals*, London, Methuen.

Thompson, I. J. (1988), 'Real Dispositions in the Physical World', *British Journal for the Philosophy of Science*, 39: 67–79.

Tooley, M. (1987), *Causation*, Oxford, Clarendon Press.

Tuomela, R. (ed.) (1978), *Dispositions*, Dordrecht, Reidel.

Weissman, D. (1965), *Dispositional Properties*, Carbondale, Southern Illinois University Press.

—— (1978), 'Dispositions as Geometrical-Structural Properties', *Review of Metaphysics*, 32: 275–97.

Wittgenstein, L. (1958), *Blue and Brown Books*, Oxford, Blackwell.

—— (1980), *Remarks on the Philosophy of Psychology*, i, G. E. M. Anscombe and G. H. von Wright (eds.), Oxford, Blackwell.

Woolhouse, R. S. (1973), 'Counterfactuals, Dispositions and Capacities', *Mind*, 82: 557–65.

Wright, A. (1990/1), 'Dispositions, Anti-Realism and Empiricism', *Proceedings of the Aristotelian Society*, 91: 39–59.

Wright, C. (1992), *Truth and Objectivity*, Cambridge, Mass., Harvard University Press.

—— (1993), *Realism, Meaning and Truth*, 2nd edn., Oxford, Blackwell.

Wright, L. (1973), 'Functions', *Philosophical Review*, 82: 139–68.

INDEX

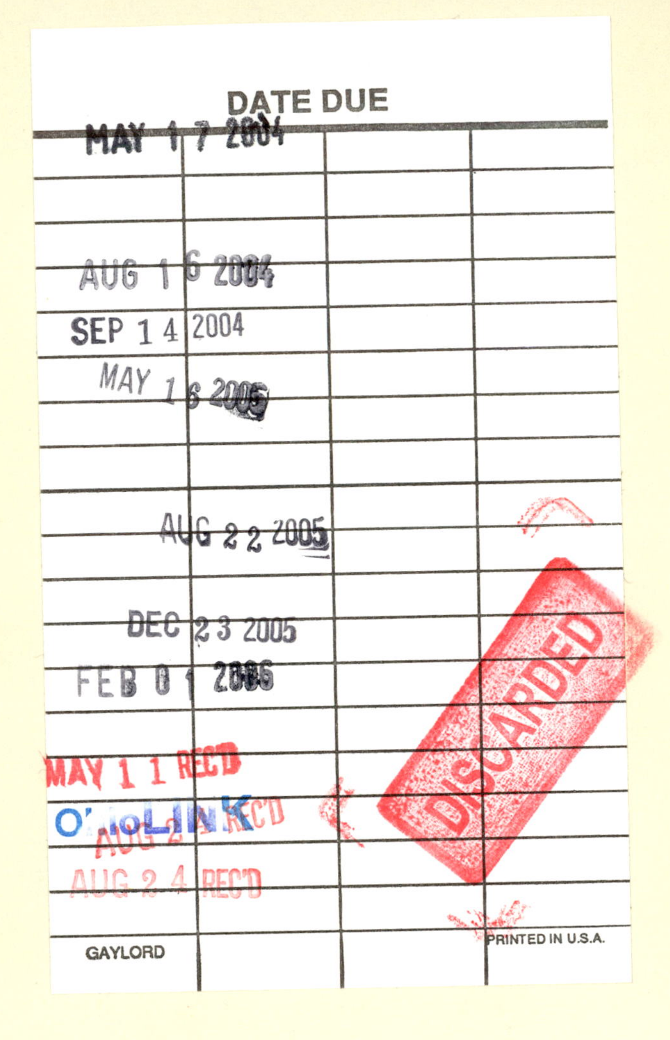
DATE DUE
MAY 17 2004
AUG 16 2004
SEP 14 2004
MAY 16 2005
AUG 22 2005
DEC 23 2005
FEB 01 2006
MAY 11 REC'D
OhioLINK
AUG 24 REC'D
DISCARDED
GAYLORD
PRINTED IN U.S.A.